UNIVERSITÄT DER BUNDESWEHR MÜNCHEN
Fakultät für Elektrotechnik und Informationstechnik

Optimization with Ruled Surface

Yayun Zhou

Vorsitzender des Promotionsausschusses: Univ.-Prof.Dr.-Ing. Rainer Marquardt
1. Berichterstatter: Univ.-Prof.Dr.rer.nat.
                              Dr.rer.nat.habil.
                              Dr.-Ing. Stefan Schäffler
2. Berichterstatter: Univ.-Prof.Dr.rer.nat.habil. Jörg Schulze

Tag der Prüfung: 12.7.2010

Mit der Promotion erlangter akademischer Grad:
Doktor-Ingenieur
(Dr.-Ing.)

Neubiberg, den 23. August 2010

Bibliografische Information der Deutschen Nationalbibliothek

Die Deutsche Nationalbibliothek verzeichnet diese Publikation in der
Deutschen Nationalbibliografie; detaillierte bibliografische Daten sind
im Internet über http://dnb.d-nb.de abrufbar.

ISBN 978-3-8325-2616-0

Logos Verlag Berlin GmbH
Comeniushof, Gubener Str. 47,
10243 Berlin
Tel.: +49 (0)30 42 85 10 90
Fax: +49 (0)30 42 85 10 92
INTERNET: http://www.logos-verlag.de

To my beloved parents and Guangxia

Acknowledgements

First of all, I would like to thank my doctor father, Professor Stefan Schäffler, for offering me such an interesting topic and his invaluable guidance and help during my PhD time. I would also like to thank my advisor in Siemens, Professor Jörg Schulze, for all his brilliant ideas and coordination. The support from these two professors made this work possible.

I owe a long time "thanks" to my colleagues in CT PP2, especially Dr. Utz Wever, Dr. Meinhard Paffrath, Dr. Inga Anita Fischer, Dr. Martin Prescher, Dr. Francesco Montrone, Dr. Efrossini Tsouchnika and Mrs. Angelika Stiller. They offer me a lot of help in work as well as in life. Their friendship and kindness are greatly appreciated.

My special thanks goes to Professor Albert Gilg, for his patience and encouragement during my stay in Siemens. He is a nice boss and a considerate helper.

I would like to thank Siemens AG CT PP2, for the financial support and all resources for my research. I would like to thank Universität der Bundewehr München, for hosting me as a PhD student.

Besides, I want to express my gratitude to all my teachers. All the knowledges and skills inherited from them are the basis for this work. I also want to thank all my friends in München and worldwide. Your accompany makes my academic journey colorful and interesting.

Finally, I want to thank my parents and my husband Guangxia, for encouraging me to pursue my dream. Their unselfish love provides me courage and strength to complete this work.

Contents

List of Figures

List of Tables

Abstract

This dissertation provides a novel design approach with respect to ruled surface, which is a special type of surface generated by moving a line in the space. Ruled surface is a favorable choice for manufacturing and can be found in many application fields.

In this dissertation, theoretical basis of ruled surface is studied. The definition of ruled surface in line geometry is combined with screw theory, algebra of dual number and kinematics. By employing the Klein mapping and the Study mapping, a ruled surface in Euclidean space is represented as a curve on a dual unit sphere (DUS). Based on this relationship, a weighted average on the dual unit sphere is defined as a minimization problem. The uniqueness of this definition is proven. It leads to a novel definition of dual spherical splines.

A series of algorithms calculating the weighted average on the dual unit sphere and interpolating the dual spherical spline are developed. Then, a complete kinematic ruled surface approximation algorithm is proposed and tested with real turbocharger blade data.

More generally, the dual spherical spline interpolation algorithm is extended to be a dual spherical spline approximation algorithm. A ruled surface is defined by several control points of a dual spherical spline. It provides an initial prototype for the blade geometry optimization with ruled surface.

Finally, combining the kinematic ruled surface approximation algorithm with the offset theory, a novel design and manufacturing strategy is proposed. A desired surface is presented as a tool path of the flank milling method with a cylindrical tool in 5-axis CNC machining. It integrates the manufacturing requirements in the design phase, which can reduce the design-cycle time and save the manufacturing cost.

Kurzfassung

Die vorliegende Dissertation stellt ein neuartiges Verfahren zur Berechnung von die Regelflächen vor. Eine Regelfläche, die durch stetige Bewegungen einer Gerade im Raum entsteht, lässt sich fertigungstechnisch besonders günstig herstellen: Das Fräsen von Oberflächen, die durch Regelflächen beschrieben werden, kann z.b. durch Flankenfräsen erheblich beschleunigt werden.

In der vorliegenden Arbeit werden zunächst die theoretischen Grundlagen der Regelfläche vorgestellt. Die Definition der Regelfläche in der Liniengeometrie wird verbunden mit der Algebra der dualen Zahlen und der Kinematik. Durch Benutzen der Klein' schen Abbildung ist eine Regelfläche im Euklidischen Raum äquivalent zu einer Kurve auf der dualen Einheitskugel. Auf Grund der Beziehung wird ein gewichteter Durchschnitt auf der dualen Einheitskugel als ein Minimierungsproblem definiert. Die Eindeutigkeit der Definition wird sichergestellt. Diese Definition führt auf eine neue Definition von sphärischen Splines auf der dualen Einheitskugel.

Zur Berechnung des gewichteten Durchschnitts und der Spline Interpolierung auf der dualen Einheitskugel werden eine Reihe von Algorithmen entwickelt.

Schliesslich wird ein kompletter Algorithmus zur kinematischen Approximierung von Regelflächen vorgestellt. Der Algorithmus wird an Hand der Schaufeln von Turboladern getestet.

Weiterhin wird der Algorithmus zur sphärischen Spline Interpolierung auf der dualen Einheitskugel weiterentwickelt zur einem Approximationsalgorithmus auf der dualen Einheitskugel. Eine Regelfläche wird damit von den Kontrollpunkten des Splines auf der dualen Einheitskugel festgelegt. Für die Schaufeln von Turboladern lassen sich damit diejenige Regelfläche finden, die eine vorhandene Schaufel am besten approximiert.

Schliesslich wird eine neue Fertigungsstrategie vorgestellt für Oberflächen, die durch Regelflächen beschrieben werden können. Dabei werden die Offsettheorie und die sphärische, kinematische Approximierung kombiniert. Die gewünschte Fläche wird dargestellt als ein Bewegungspfad für den zylinderfömigen Flankenkopf einer 5-Achsen CNC-Maschine. Durch das vorgestellte Verfahren werden Fertigungsanforderungen schon in der Design-Phase berücksichtigt, was zu Kostenvorteilen und zu schnelleren Designzyklen führen kann.

Chapter 1
Introduction

The development of Computer Aided Geometric Design (CAGD) provides engineers a new process to design a product. With the help of CAGD systems, a designer can translate a set of functional requirements into a 3D computer model and optimize the model before manufacturing. Today's engineers can use CAGD systems to design a wide variety of surface types which meet the efficiency and stability requirements. However, some of them may be expensive to produce and there is no guarantee that those surfaces can be produced accurately. In many applications, like power, automotive and aerospace applications, manufacturability and manufacturing cost are important criteria for the geometric design.

Nowadays, there are various manufacturing methods. Some still use the traditional tools, such as mills [57] and lathes [32]; others employ new technologies, such as EDM (Electrical Discharge Machining) [10] [35], ECM (Electro Chemical Machining)[49], laser cutting [31], and etc. Those tools are usually controlled by a Computer Numerical Control (CNC) system [58], which is the most common Computer Aided Manufacturing (CAM) system. The tool trajectories can be generated from the computer models, then a computer "controller" drives the tool accordingly to fabricate components by selectively removal of material. The CNC system creates a physical realization of the computer model.

1.1 Motivation

The purpose of this dissertation is to provide a robust shape design approach which is easy and cheap to manufacture. Among all the surfaces, ruled surface is a favorable choice in manufacturing. It is a special type of surface which can be generated by moving a line in the space. They occur in several applications such as wire EDM and laser cutting, which treat the cutting tool as a moving line. Besides, it is well known that a ruled surface can be efficiently produced using the side milling method with a cylindrical cutter in the CNC machining. Hence, the manufacturing cost of producing a ruled surface is much less than a non-ruled surface, which is appealing in industry.

In mathematics, a surface Φ is ruled if through every point of Φ there is a straight line that lies on Φ. Ruled surface is one of the simplest objects in geometric modeling. In real world, many surfaces are ruled surfaces. Plane, cylinder and cone are some examples of ruled surfaces that we are familiar with. Besides, hyperbolic paraboloid and hyperboloid are also special ruled surfaces. In Euclidean space $\mathbb{R}^3$, a ruled surface Φ possesses a parametric representation [15]:

$$\mathbf{x}(u,v) = \mathbf{a}(u) + v\mathbf{r}(u), \quad u \in I, v \in \mathbb{R}, \tag{1.1}$$

where $\mathbf{a}(u)$ is called the *directrix curve* and $\mathbf{r}(u)$ is the direction vector of the *generator*.

Alternatively, a ruled surface Φ can be parameterized by two directrix curves $\mathbf{p}(u)$ and $\mathbf{q}(u)$:

$$\mathbf{x}(u,v) = (1-v)\mathbf{p}(u) + v\mathbf{q}(u). \tag{1.2}$$

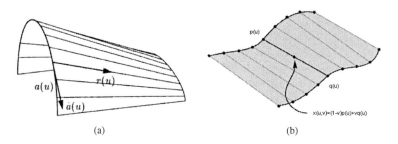

(a)   (b)

Fig. 1.1 Representations of ruled surface: (a) Representation I; (b) Representation II.

The straight line denoted as $\mathbf{x}(u_0,v) = (1-v)\mathbf{p}(u_0) + v\mathbf{q}(u_0)$ is called a *ruling*. The Gaussian curvature on a ruled regular surface is everywhere non-positive. In particular, if the Gaussian curvature of a ruled surface is zero, this type of ruled surface is called a *developable surface*. It is a "surface" which can be flattened onto a plane without distortion. Conversely, a developable surface is a surface which can be made by transforming a plane.

Although ruled surfaces have been studied extensively in classical geometry, they have not been fully exploited for applications in geometric design and manufacturing. In 1991, Ravani and Wang first applied concepts of Bézier curve and surface design to construct the ruled surface [46]. Followed by Hoschek, Schwanecke, Pottmann and Peternell, the properties of ruled surface in line geometry are densely studied[38] [43] [44]. Some ruled surface interpolation and approximation algorithms are developed based on line geometry [20] [6] [41] [36][37]. The most recent work of ruled surface approximation is inspired by the level set method, employing the second order approximation of the squared distance function as an error measurement to avoid the parametrization problem [42] [41]. Using those algorithms, an arbitrary given surface or scattered points can be approximated by a ruled surface.

In recent years, the line geometry is used in kinematics together with the screw theory to describe the geometric properties of the screw axis of a moving rigid body, which describes the manufacturing procedure of a ruled surface. However, the focus of this research is to generate a ruled surface kinematically [52][53] [55] [33] [64]. Normally, the kinematically generated ruled surface does not pass the control line structure. The geometry design part and manufacturing part are separated: in the design stage, a given surface is approximated with a ruled surface; then in the manufacturing stage, the ruled surface is generated kinematically to guide the manufacturing [62]. Therefore, the kinematic approximation of ruled surface has high demand in industrial manufacturing and has many applications in different fields, especially in fluid machines, such as propellers, impellers of centrifugal compressors, rotors of turbo molecular pumps, gas turbine and turbocharger.

1.2 Work in the Dissertation

In this dissertation, a series of ruled surface design and optimization approaches are presented. The essence is a novel kinematic ruled surface approximation algorithm. This algorithm is developed based on the theory of line geometry and screw theory. In screw theory and dual vector algebra, there is a one-to-one correspondence between an oriented straight line in 3-dimensional Euclidean space and a point on the surface of a dual unit sphere (DUS). Using the Klein map and the Study map, a ruled surface in Euclidean space is transformed to a curve on the DUS. Then the approximation/interpolation of ruled surface in Euclidean space can be converted to a curve approximation/interpolation problem on the DUS. Inspired by the definition of weighted average in the real sphere space, we first define a weighted average in the DUS space. The weighted average in the DUS space is defined as a result of a least squares minimization problem, which allows a novel method of defining Bézier and spline curves on the DUS. We prove that the least squares minimization problem has a unique solution if the input points locate on a hemisphere. The continuity and convexity properties of the dual spherical spline are also discussed. Based on these definitions, a kinematic ruled surface approximation algorithm is developed. The ruled surface is represented as a dual spherical spline, which denotes the trajectory of the moving ruling. The kernel of this algorithm is a fast algorithm of dual spherical spline interpolation on the DUS. This spherical spline allows the use of arbitrary knot positions and the algorithm is fast enough for real time application. Combining the offset theory, this algorithm can be extended to control the manufacturing process.

The organization of the dissertation is as follows. Chapter 2 introduces the theoretical background related to the topics. It includes basic concepts in line geometry, curve and surface definition in CAGD, screw theory, algebra of dual number and kinematics. Chapter 3 reviews the related work in ruled surface approximation and implements a typical algorithm to approximate a given blade surface. Chapter 4 defines a weighted average on the DUS and the corresponding dual spherical spline. In addition, the uniqueness of the definition is proven and the continuity and convexity properties of the dual spherical splines are discussed. Chapter 5 first introduces an algorithm calculating a weighted average on the DUS, followed by an algorithm of dual spherical spline interpolation. In Chapter 6, a complete kinematic ruled surface approximation algorithm is proposed, it is tested using the real blade data and can be used to guide the wire EDM and the laser cutting, which treat the cutting tool as a moving line. Chapter 7 extends the dual spherical spline interpolation algorithm to a dual spherical spline approximation algorithm. Combined with a rule surface and free-form surface intersection algorithm [60], it provides an initial prototype for a blade geometry optimization with ruled surface. Chapter 8 investigates the side milling process using cylindrical tools in CNC machining and applies the kinematic approximation algorithm of ruled surface into CNC interpolation to generate a tool path. Chapter 9 evaluates the work and proposes the future research.

Chapter 2
Theoretical Analysis of Ruled Surface

In this chapter, some fundamental concepts related to the topics in this dissertation are introduced. It contains two major parts, line geometry and kinematics. First, we set up the concept of line geometry. The projective space P^n is defined based on the extension of Euclidean space. The algebraic representation of a line is discussed and linked to a point in P^5 through the Klein mapping. The ruled surface is defined based on the line geometry. We also introduce the definitions of Bézier curve and surface in this part, which lead to useful representations of rule surface in CAGD. Many ruled surface approximation algorithms, which will be discussed in Chapter 3, are developed based on line geometry. In the second part of this chapter, some useful concepts of kinematics are introduced, including the screw theory and the algebra of dual numbers. Finally, dual vector representations of a line and a ruled surface are presented. Those ruled surface representations are essential for the kinematic rule surface approximation algorithms introduced in the following chapters.

2.1 Line Geometry

The theory of line geometry was first introduced by Plücker in his work published in 1865 [40]. Other mathematicians, like Klein and Study, made contributions on this field. The most recent book related in this field is published by Pottmann [44], which provides an overview of this theory.

In this part, some important concepts in line geometry are introduced. First, the Euclidean space is extended to the projective space. Curves and surfaces are also considered from the viewpoint of the projective geometry. Then, several mathematical tools for rational curves and surfaces are discussed. Finally, an algebraic ruled surface representation is introduced.

2.1.1 Projective space

The *projective space* is nothing but a projective extension of the Euclidean space. Adding a point at infinity to each line L in the real Euclidean plane, we get a real projective plane. Those points at infinity are called *ideal points* and the points related to the Euclidean plane are called *proper points*. Thus, two parallel lines share the same ideal point in projective plane. It has a one-to-one correspondence between one-dimensional linear subspaces of $\mathbb{R}^3$ and points of the projectively extended plane. The coordinate of a point in the projective plane is:

$$\mathbf{x}\mathbb{R} = (\lambda, \lambda x, \lambda y) = (x_0 : x_1 : x_2), \ \lambda \in \mathbb{R}. \tag{2.1}$$

When $x_0 = 0$, the coordinate denotes an ideal point. When $x_0 \neq 0$, the coordinate stands for a proper point, which can be transformed back to the Cartesian coordinate system of the Euclidean plane by:

$$(x_0, x_1, x_2)\mathbb{R} \in P^2 \iff (x_1/x_0, x_2/x_0) \in \mathbb{R}^2, \ (x_0 \neq 0). \tag{2.2}$$

The construction of an n-dimensional real projective space P^n is completely analogous to that of P^2. P^n is a set of one-dimensional subspaces of $\mathbb{R}^{n+1}$. The homogeneous coordinate for a point in P^n is:

$$\mathbf{x}\mathbb{R} = (x_0 : x_1 : \ldots : x_n). \tag{2.3}$$

When $x_0 = 0$, it is an ideal point. Otherwise, $\mathbf{x}\mathbb{R}$ is a proper point, and its coordinate in $\mathbb{R}^n$ are recovered by:

$$(x_0, \ldots, x_n)\mathbb{R} \in P^n \iff (x_1/x_0, \ldots, x_n/x_0) \in \mathbb{R}^n, \ (x_0 \neq 0). \tag{2.4}$$

For more information about the projective space, please refer to [44].

2.1.2 Line space model

Under the homogeneous Cartesian coordinate system, a line L can be represented algebraically by two distinct points $X = \mathbf{x}\mathbb{R}$ and $Y = \mathbf{y}\mathbb{R}$ on the line:

$$\mathbf{l}(t) = (1-t)\mathbf{x} + t\mathbf{y}. \tag{2.5}$$

Equivalently, a line L in P^3 can be represented by the exterior product of two points $X \wedge Y$, which is called the *homogeneous Plücker vector coordinate* $(\mathbf{l}, \bar{\mathbf{l}})$: [44]

$$(x_0, \ldots, x_3) \wedge (y_0, \ldots, y_3) = (l_{01} : l_{02} : l_{03} : l_{23} : l_{31} : l_{12}), \ l_{ij} = x_i y_j - x_j y_i. \tag{2.6}$$

If we write $\mathbf{X} = (x_0, x_1, x_2, x_3)\mathbb{R} = (x_0, \mathbf{x})\mathbb{R}$ and $\mathbf{Y} = (y_0, y_1, y_2, y_3)\mathbb{R} = (y_0, \mathbf{y})\mathbb{R}$, the homogeneous Plücker coordinate can be written in the form:

$$L = (l_{01} : l_{02} : l_{03} : l_{23} : l_{31} : l_{12}) = (x_0\mathbf{y} - y_0\mathbf{x}, \mathbf{x} \times \mathbf{y}) = (\mathbf{l}, \bar{\mathbf{l}}), \tag{2.7}$$

from which we can see that these coordinate elements are not independent, they fulfill the Plücker relation:

$$\Omega_q(L) = \mathbf{l} \cdot \bar{\mathbf{l}} = l_{01}l_{23} + l_{02}l_{31} + l_{03}l_{12} = 0. \tag{2.8}$$

The homogeneous Plücker coordinate $\mathbf{L} = (l_{01} : l_{02} : l_{03} : l_{23} : l_{31} : l_{12})$ defines a point $L\mathbb{R}$ in P^5. We need to notice that not every point in P^5 is a Plücker coordinate. Only the points satisfying the Plücker relation Eq. (2.8) are Plücker coordinates. Eq. (2.8) defines a quadratic variety in P^5, called *Klein quadric* M_2^4. The upper index denotes the dimension and the lower index denotes the degree of the variety. Thus we can set up the bijection mapping $\gamma : L \mapsto M_2^4$ between lines $L \in P^3$ and points $L\mathbb{R}$ of M_2^4. This mapping is called *Klein mapping*.

2.1.3 Curves and surfaces in geometric design

Now, we can define curves and surfaces in the projective space [44]. A *curve* in P^n can be defined as a mapping $c: I \mapsto P^n$, where I is some parameter interval, and

$$c(t) = \mathbf{c}(t)\mathbb{R}, \text{ with } \mathbf{c}(t) = (c_0(t), c_1(t), \dots, c_n(t)). \tag{2.9}$$

Closed curves can be defined either as smooth mappings of a unit circle into the projective space or as periodic mappings of a real line. Similarly, an *m-surface* in P^n can be defined as a mapping $s: D \mapsto P^n$ where $D \subset \mathbb{R}^m$, and

$$s(\mathbf{u}) = \mathbf{s}(\mathbf{u})\mathbb{R} = (s_0(\mathbf{u}), \dots, s_n(\mathbf{u}))\mathbb{R}, \ \mathbf{u} = (u_1, \dots, u_m) \in D. \tag{2.10}$$

The definition of Bézier curve and B-spline curve can also be extended to the projective space. The Bézier curve $c: \mathbb{R} \mapsto P^m$ with $c(t) = \mathbf{c}(t)\mathbb{R} = (\sum_{i=0}^n B_i^n(t)\mathbf{b}_i)\mathbb{R}$, ($\mathbf{b}_i \in \mathbb{R}^{m+1}$) is called a *rational Bézier curve*. The Bernstein polynomials of degree n are given by:

$$B_i^n(t) = \binom{n}{i} t^i (1-t)^{n-i}, \ (i = 0, \dots, n). \tag{2.11}$$

The points $B_i = \mathbf{b}_i\mathbb{R}$ are called *control points* and the points $F_i = (\mathbf{b}_i + \mathbf{b}_{i+1})\mathbb{R}$ are called *frame points*. When the control points and frame points are given, we can construct a curve $c(t)$ by the De Casteljau algorithm [16].

Similarly, we can write a *rational B-spline curve* $s(u)$ in P^d with control points $\mathbf{d}_0, \dots, \mathbf{d_m}$ and knot vector $\mathbf{T}$ as:

$$s(u) = \mathbf{s}(u)\mathbb{R} = \left(\sum_{i=0}^{m} N_i^n(u)\mathbf{d}_i\right)\mathbb{R}, \tag{2.12}$$

where $N_i^n(u)$ is the i-th B-spline basis function of degree n corresponding to the knot vector $\mathbf{T} = (t_0 \leq t_1 \leq \ldots \leq t_{m+n+1})$:

$$
\begin{aligned}
N_i^0 &:= \begin{cases} 1 & \text{for } t_i \leq u \leq t_{i+1} \\ 0 & \text{else} \end{cases}, \\
N_i^r &:= \frac{u-t_i}{t_{i+r}-t_i}N_i^{r-1}(u) + \frac{t_{i+r+1}-u}{t_{i+r+1}-t_{i+1}}N_{i+1}^{r-1}(u).
\end{aligned}
\tag{2.13}
$$

It is also called *NURBS* (Non-Uniform Rational B-Spline). Its control points $\mathbf{d}_i\mathbb{R}$ and frame points $\mathbf{f}_i\mathbb{R} = (\mathbf{d}_i + \mathbf{d}_{i+1})\mathbb{R}$ constitute the geometric control polygon of the curve.

A *rational Bézier surface* is a surface of the form:

$$f(x_1,\ldots,x_d) = \mathbf{f}(x_1,\ldots,x_d)\mathbb{R}, \tag{2.14}$$

where $\mathbf{f}$ is an ordinary polynomial Bézier surface. If $\mathbf{f}$ is a polynomial B-spline surface, Eq. (2.14) stands for a *rational B-spline surface*. Rational tensor product B-spline surfaces are also called *NURBS surfaces* [44].

Using the Bézier basis function and the B-spline basis function, we can define a *tensor product surface*. The tensor product surface is a generalization of the notion of Bézier and B-spline curve. A tensor product 2-surface with degree (n,m) is given:

$$s(u,v) = \sum_{i=0}^{n}\sum_{j=0}^{m} B_i^n(u)B_j^m(v)\mathbf{b}_{ij}, \tag{2.15}$$

where $\mathbf{b}_{ij}$ are control points. Replacing the Bernstein polynomials in Eq. (2.15) by B-spline basis functions, we get a *tensor product B-spline surface*:

$$s(u,v) = \sum_{i=0}^{n}\sum_{j=0}^{m} N_i^n(u)N_j^m(v)\mathbf{d}_{ij}. \tag{2.16}$$

2.1.4 Algebraic representation of ruled surface

2.1.4.1 Ruled surface in P^3-space

In the introduction, we have seen the algebraic ruled surface representation in Euclidean space. In this section, we define the ruled surface in the projective space and study its properties.

The ruled surface is a one-parameter family of lines in P^3. Ruled surfaces can be redefined using the Klein mapping γ [44]:

definition 2.1.1 *A family $\mathscr{R} = \mathbf{R}(u)$ of lines in P^3 is called a ruled surface, if its Klein image $\mathscr{R}\gamma$ is a curve $\mathbf{R}\gamma(u)$ in the Klein quadric M_2^4. Its differentiability class is defined to be the differentiability class of $\mathscr{R}\gamma$. The lines $\mathbf{R}(u)$ are the generator lines (rulings) of the surface.*

Obviously, a ruled surface is also a two-parameter set of points in P^3. One can prove that it is possible to span each ruling $\mathbf{R}(u)$ by two points running on C^r parametric curves: $\mathbf{a}_0(u)\mathbb{R}$ and $\mathbf{a}_1(u)\mathbb{R}$, which are known as *directrix curves* [38]. The ruled surface as a point set $\Phi \subset P^3$ is parameterized over $I \times P^1$:

$$\mathbf{x}(u,\lambda) = \lambda_0 \mathbf{a_0}(u) + \lambda_1 \mathbf{a_1}(u) \text{ with } \lambda = (\lambda_0, \lambda_1) \in P^1. \tag{2.17}$$

Equivalently, we can define the ruled surface in Plücker coordinates:

$$\mathbf{R}(u) = \mathbf{a_0}(u) \wedge \mathbf{a_1}(u). \tag{2.18}$$

A generator line $\mathbf{R}(u_0)$ of a C^1 ruled surface is called regular if the Klein image $\mathbf{R}(u_0)\gamma$ is a regular point of the curve $\mathbf{R}\gamma(u)$. If $\mathbf{R}(u) = (\mathbf{r}(u), \bar{\mathbf{r}}(u))\mathbb{R}$, regular generator line $\mathbf{R}(u_0)$ means linear independence of $(\mathbf{r}(u_0), \bar{\mathbf{r}}(u_0))$ and $(\dot{\mathbf{r}}(u_0), \dot{\bar{\mathbf{r}}}(u_0))$. Their span $\mathbf{R}\gamma(u_0) \vee \dot{\mathbf{R}}\gamma(u_0)$ is the tangent of the curve $\mathbf{R}\gamma$ at the parameter value u_0.

A regular generator $\mathbf{R}(u_0)$ of a C^1 ruled surface $\mathscr{R}$ is called *torsal* if the tangent $T(u_0)$ of the Klein image $\mathbf{R}\gamma(u)$ at $\mathbf{R}(u_0)\gamma$ is contained in the Klein quadric M_2^4, otherwise, the regular generator is called *non-torsal*. The singular surface point of a torsal generator is called its *cuspidal point*, and the tangent plane in its other points is called *torsal plane*. Using the representation in Eq. (2.18), we get the following results:

$$\begin{aligned} rank(\mathbf{a}_0(u_0), \dot{\mathbf{a}}_0(u_0), \mathbf{a}_1(u_0), \dot{\mathbf{a}}_1(u_0)) = 2 &\Leftrightarrow \mathbf{R}(u_0) \text{ is singular,} \\ rank(\mathbf{a}_0(u_0), \dot{\mathbf{a}}_0(u_0), \mathbf{a}_1(u_0), \dot{\mathbf{a}}_1(u_0)) = 3 &\Leftrightarrow \mathbf{R}(u_0) \text{ is torsal,} \\ rank(\mathbf{a}_0(u_0), \dot{\mathbf{a}}_0(u_0), \mathbf{a}_1(u_0), \dot{\mathbf{a}}_1(u_0)) = 4 &\Leftrightarrow \mathbf{R}(u_0) \text{ is non-torsal.} \end{aligned} \tag{2.19}$$

In other words, if the tangents of the directrix curves $\mathbf{a}_1(u_0)$ and $\mathbf{a}_2(u_0)$ are contained in the same line, the generator line (ruling) is singular; if they span a plane, then the ruling is regular and torasl; if they are screw and span entire P^3, the ruling is non-torsal. If a ruled surface whose rulings are all torsal, it is called a *torsal ruled surface* (see Fig. 2.1). If a ruled surface is not torsal, it is called a *skew ruled surface*. In the following discussion, we will focus on the skew ruled surface.

If the Klein image curves $\mathbf{R}_1\gamma(u)$, $\mathbf{R}_2\gamma(u)$ of two ruled surfaces $\mathbf{R}_1(u)$ and $\mathbf{R}_1(u)$ possess contact of order r at a regular point $\mathbf{R}(u_0)\gamma$, then the surface $\mathbf{R}_1(u)$ and $\mathbf{R}_2(u)$ possess contact of order r along the common regular generator $\mathbf{R}(u_0)$ [38].

2.1.4.2 Rational ruled surfaces and their Bézier representation

A real rational ruled surface $\mathscr{R}$ possesses a real rational Klein image curve $\mathscr{R}\gamma$. This curve has a real parametrization $\mathbf{R}(u)\mathbb{R}$, which is polynomial in a homogeneous

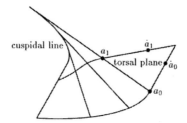

Fig. 2.1 Torsal ruled surface.

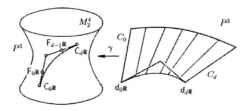

Fig. 2.2 Rational ruled surface and its Klein image curve.

parameter u. The Bézier representation of the Klein image is:

$$\mathbf{R}(u) = \sum_{i=0}^{n} B_i^n(u)\mathbf{C}_i \text{ with } u = (u_0 : u_1) \in P^1, \qquad (2.20)$$

where the Bernstein basis functions are:

$$B_i^n(u) := \binom{n}{i} u_1^i (u_0 - u_1)^{n-i}. \qquad (2.21)$$

The $n+1$ points $\mathbf{C_i}\mathbb{R} \in P^5$ are Bézier control points, and $\mathbf{F}_i\mathbb{R} = (\mathbf{C}_i + \mathbf{C}_{i+1})\mathbb{R}$ are its frame points.

The Bézier representation given by Eq. (2.20) is useful to compute the planar intersection and the contour of ruled surface. For details, please refer to [38].

Furthermore, a ruled surface can also be written as a *rational tensor product surface* of degree $(d, 1)$:

$$\mathbf{x}(u,v) = (1-v)\sum_{i=0}^{d} B_i^d(u)\mathbf{b}_{i0} + v\sum_{i=0}^{d} B_i^d(u)\mathbf{b}_{i1} = \sum_{i=0}^{d}\sum_{j=0}^{1} B_i^d B_j^1 \mathbf{b}_{ij}. \qquad (2.22)$$

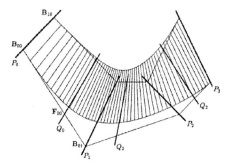

Fig. 2.3 Line geometric control structure of ruled surface.

This is the standard CAGD representation of a ruled surface, from which we can get the line geometric control structure. The control lines P_0, $P_1, \cdots, P_d$ are connections between control points B_{i0} and B_{i1}. The frame lines Q_0, $Q_1, \cdots, Q_d$ are connections between frame points F_{i0} and F_{i1}. The line geometric control structure is convenient for the interactive design of ruled surfaces, when we do not want to put special emphasis on boundary curves. The representation given by Eq. (2.22) can convert to Eq. (2.20). The line coordinate representation $\mathbf{R}(u) = \mathbf{R}(u,0) \wedge \mathbf{R}(u,1)$ is:

$$\mathbf{R}(u) = \sum_{k=0}^{2d} B_k^{2d}(u)\mathbf{C}_k \text{ with } \mathbf{C}_k = \frac{1}{\binom{2d}{k}} \sum_{i+j=k} \binom{d}{k}\binom{d}{j} \mathbf{b}_{i0} \wedge \mathbf{b}_{i1}. \qquad (2.23)$$

Notice that the degree of $\mathbf{R}$ is in general $2d$. The line geometric control structure of a rational rule surface is shown in Fig. 2.3

Eq. (2.22) describes a surface as a focus of points, a ruled surface can also be interpreted as an envelope of its tangent planes. Due to the principal of duality in projective geometry, we can change "points" to "planes", then a representation of a *dual tensor product Bézier surface* is obtained:

$$\begin{aligned}\mathbf{x}(u,v) &= (1-v)\sum_{i=0}^d B_i^d(u)\mathbf{U}_{i0} + v\sum_{i=0}^d B_i^d(u)\mathbf{U}_{i1} \\ &:= (x_0(u,v),x_1(u,v),x_2(u,v),x_3(u,v)),\end{aligned} \qquad (2.24)$$

where the vectors $\mathbf{U}_{ij}$ are the homogeneous plane coordinate vectors of the control plane. Eq. (2.24) describes a two parametric set of planes whose explicit equation in $\mathbb{R}^3$ is:

$$x_0(u,v) + x_1(u,v)x + x_2(u,v)y + x_3(u,v)z = 0. \tag{2.25}$$

The required ruled surface is the envelope of the planes described by Eq. (2.24).

2.2 Kinematics

In this part, some concepts in kinematics, which are the basis for the kinematic ruled surface approximation algorithm, are introduced. First, the dual number is defined, later the essence of screw theory is described and linked to Plücker coordinates of lines. Finally, a ruled surface is represented adopting dual vectors.

2.2.1 Dual number

The dual numbers were first introduced by Clifford [9]. A *dual number* can be written in the form $\hat{a} = a + \varepsilon a^\circ$, where $a, a^\circ \in \mathbb{R}$ and ε is the dual element with:

$$\begin{aligned} &\varepsilon \neq 0, \\ &0\varepsilon = \varepsilon 0 = 0, \\ &1\varepsilon = \varepsilon 1 = \varepsilon, \\ &\varepsilon^2 = 0. \end{aligned} \tag{2.26}$$

Two dual numbers $\hat{a} = a + \varepsilon a^\circ$ and $\hat{b} = b + \varepsilon b^\circ$ equal, if and only if $a = b, a^\circ = b^\circ$. The addition and multiplication for two dual numbers are defined as follow:

$$\hat{a} + \hat{b} = (a + b) + \varepsilon(a^\circ + b^\circ), \tag{2.27a}$$

$$\hat{a} \cdot \hat{b} = a \cdot b + \varepsilon(a \cdot b^\circ + b \cdot a^\circ). \tag{2.27b}$$

We can see that the set of dual numbers, denoted as $\mathbb{D}$, forms a commutative group under addition. The associative law holds for multiplication and dual numbers are distributive. Thus, the set of dual numbers forms a ring over $\mathbb{R}$. Division of dual numbers is defined only when $b \neq 0$:

$$\frac{\hat{a}}{\hat{b}} = \frac{a}{b} + \varepsilon \frac{a^\circ \cdot b - a \cdot b^\circ}{b^2}. \tag{2.28}$$

Using this property, we can calculate the derivative of a dual number function $\hat{F}$. Rewrite the function $\hat{F}(x + \varepsilon x^\circ)$ as:

$$\hat{F}(x + \varepsilon x^\circ) = f(x, x^\circ) + \varepsilon g(x, x^\circ), \tag{2.29}$$

where $f(x, x^\circ)$ and $g(x, x^\circ)$ are real functions of two real variables x and x°. Writing the expression for the derivative, we have:

$$\frac{d\hat{F}}{d\hat{x}} = \frac{df(x,x^\circ) + \varepsilon dg(x,x^\circ)}{dx + \varepsilon dx^\circ}$$
$$= \left(\frac{\partial f}{\partial x} + \frac{\partial f}{\partial x^\circ}\frac{dx^\circ}{dx}\right) + \varepsilon\left[\frac{\partial g}{\partial x} + \left(\frac{\partial g}{\partial x^\circ} - \frac{\partial f}{\partial x}\right)\frac{dx^\circ}{dx} - \frac{\partial f}{\partial x^\circ}\left(\frac{dx^\circ}{dx}\right)^2\right]. \tag{2.30}$$

Similar as the functions of complex variables, the derivative has a condition to satisfy: the limit of the ratio of the increment $d\hat{F}(\hat{x})$ of the function $\hat{F}(\hat{x})$ to the increment $d\hat{x}$ of the dual number $\hat{x}$ as $d\hat{x} \rightarrow 0$ is independent of the ratio $dx^\circ : dx$. Therefore, we have:

$$\frac{\partial f}{\partial x^\circ} = 0; \quad \frac{\partial g}{\partial x^\circ} = \frac{\partial f}{\partial x}. \tag{2.31}$$

The first equation implies that the function f is a function only of the variable x, i.e., $f(x,x^\circ) = f(x)$. The second equation implies the expression for the function g:

$$g(x,x^\circ) = x^\circ\frac{\partial f}{\partial x} + f^\circ(x), \tag{2.32}$$

where $f^\circ(x)$ is a certain function of x. Therefore, we get the derivative of the function $\hat{F}(\hat{x})$:

$$\frac{d\hat{F}(\hat{x})}{d\hat{x}} = \frac{\partial f}{\partial x} + \varepsilon\frac{\partial}{\partial x}\left(x^\circ\frac{\partial f}{\partial x} + f^\circ(x)\right)$$
$$= \frac{\partial f}{\partial x} + \varepsilon\left(x^\circ\frac{\partial^2 f}{\partial x^2} + \frac{\partial f^\circ}{\partial x}\right). \tag{2.33}$$

The differentiation with respect to the dual number $\hat{x}$ is reduced to the differentiation with respect to the real variable x. We have the very important theorem in screw calculus[11]:

theorem 2.2.1 *All theorems of differential and integral calculus are preserved in the domain of dual numbers of the form $x + \varepsilon x^\circ$.*

A differentiable function $F(x)$ can be defined for a dual argument $F(x + \varepsilon x^\circ)$ by expanding the function using a Taylor series. Since $\varepsilon^2 = 0$, we get:

$$F(x + \varepsilon x^\circ) = F(x) + \varepsilon x^\circ\frac{dF}{dx}. \tag{2.34}$$

2.2.2 Screw theory

Screw theory was first appeared at the beginning of 19th century following by the papers of Poinsot, Chasles and Möbius. The work of these authors laid the foundations for the study of kinematics and statics. The notion of screw was formulated later by Plücker in his papers. The monumental work of screw theory was published by R. Ball in 1876. In 1895, screw calculus was constructed in the work of A. P. Kotel'nikov, using dual numbers to represent screws.

In the sense of rigid motion, a screw is a way of describing a displacement. It can be thought as a rotation about an axis and a translation along that same axis. A general screw $\hat{s}$ consists of two parts, a real 3-vector s which indicates the *direction* of the the screw, and a real 3-vector s_p which locates $\hat{s}$ by recording the *moment* of

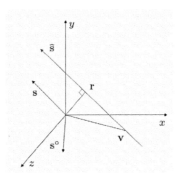

Fig. 2.4 Geometric interpretation of a screw.

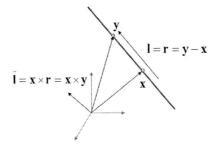

Fig. 2.5 Plücker coordinate of an oriented line in E^3.

the screw about the origin [29]. In these terms, a screw is formed as:

$$\hat{\mathbf{s}} = \mathbf{s} + \varepsilon \mathbf{s}_p = \mathbf{s} + \varepsilon(p\mathbf{s} + \mathbf{s}_0), \qquad (2.35)$$

where $\mathbf{s}_0 = \mathbf{r} \times \mathbf{s} = \mathbf{v} \times \mathbf{s}$, and in which ε is a dual element satisfying $\varepsilon^2 = 0$. Here, p is the pitch of the screw and $\mathbf{s}_0$ is the moment of the line of the screw about the origin. $\mathbf{s}_0$ is orthogonal to $\mathbf{s}$ ($\mathbf{s} \cdot \mathbf{s}_0 = 0$). The moment $\mathbf{s}_0$ is derived as shown from the origin radius vector $\mathbf{r}$, or more generally from any point $\mathbf{v}$ of the screw. The length $\|\mathbf{s}\|$ of the real part $\mathbf{s}$ is the real magnitude of the screw. Fig. 2.4 illustrates a geometric interpretation of a screw.

A line is a screw in which the pitch is zero-valued; i.e., $p = 0$. Therefore, we can see that the Plücker coordinate $(\mathbf{l}, \bar{\mathbf{l}})$ of an oriented line in $\mathbb{R}^3$ can also be interpreted as a screw. The vector $\mathbf{l}$ indicates the direction of $\mathbf{l}$, which is usually normalized to $\|\mathbf{l}\| = 1$. The vector $\bar{\mathbf{l}}$ is the moment vector with respect to the origin. The geometric interpretation of a Plücker coordinate can be shown in Fig. 2.5.

2.2.3 Dual representation of line

Dual numbers can be extended to the vector space, the space $\mathbb{D}^3$ is defined as a set of all pairs of vectors:

$$\hat{\mathbf{a}} = \mathbf{a} + \varepsilon \mathbf{a}^\circ \text{ where } \mathbf{a}, \mathbf{a}^\circ \in \mathbb{R}^3. \tag{2.36}$$

Given two dual vectors, $\hat{\mathbf{x}} = \mathbf{x} + \varepsilon \mathbf{x}^\circ$ and $\hat{\mathbf{y}} = \mathbf{y} + \varepsilon \mathbf{y}^\circ$, we can define the inner product and cross product in $\mathbb{D}^3$:

$$\hat{\mathbf{x}} \cdot \hat{\mathbf{y}} = \mathbf{x} \cdot \mathbf{y} + \varepsilon(\mathbf{x}^\circ \cdot \mathbf{y} + \mathbf{x} \cdot \mathbf{y}^\circ), \tag{2.37a}$$

$$\hat{\mathbf{x}} \times \hat{\mathbf{y}} = (\mathbf{x} \times \mathbf{y}) + \varepsilon((\mathbf{x}^\circ \times \mathbf{y}) + (\mathbf{x} \times \mathbf{y}^\circ)). \tag{2.37b}$$

The length of a dual vector is defined as

$$\|\hat{\mathbf{x}}\| = \mathbf{x} \cdot \mathbf{x}^{\frac{1}{2}} + \varepsilon \frac{\mathbf{x} \cdot \mathbf{x}^\circ}{\mathbf{x} \cdot \mathbf{x}}. \tag{2.38}$$

A dual vector with length 1 is called a *dual unit vector*. Obviously, a dual unit vector satisfies:

$$\mathbf{x} \cdot \mathbf{x} = 1, \tag{2.39a}$$

$$\mathbf{x} \cdot \mathbf{x}^\circ = 0. \tag{2.39b}$$

The dual unit vector can be used to represent a line. Recall from the previous section, the Plücker coordinate $(\mathbf{l}, \bar{\mathbf{l}})$ satisfies the relationship:

$$\mathbf{l} \cdot \mathbf{l} = 1, \tag{2.40a}$$

$$\mathbf{l} \cdot \bar{\mathbf{l}} = 0. \tag{2.40b}$$

Then, a more compact way to represent a line is derived: the dual vector representation of a line is simply writing the Plücker coordinate as a dual unit vector.

Dual unit vectors define points on a sphere in $\mathbb{D}^3$. This sphere is refereed as the *dual unit sphere* (DUS). The computational problem of computing points on a quadric in P^5 is reduced to a problem in a dual form of spherical geometry [55]. This leads to a new interpretation of the inner product and cross product of two lines. If we define the dual angle between two lines as $\hat{\theta} = \theta + \varepsilon d$, where θ is the angel between the two lines and d is the minimum distance between two lines. Then the inner product of two lines $\hat{\mathbf{x}}$ and $\hat{\mathbf{y}}$ can be written as:

$$\hat{\mathbf{x}} \cdot \hat{\mathbf{y}} = \cos \hat{\theta}. \tag{2.41}$$

The cross product can be written as:

$$\hat{\mathbf{x}} \times \hat{\mathbf{y}} = \hat{\mathbf{n}} \sin \hat{\theta}, \tag{2.42}$$

where $\hat{\mathbf{n}}$ is a line through the common perpendicular between the two lines.

2.2.4 Dual vector representation of ruled surface

Since lines in $\mathbb{R}^3$ correspond to points on the DUS, a ruled surface in $\mathbb{R}^3$ relates to a curve on the DUS. In this form, a ruled surface can be written as:

$$\hat{L}(u) = \mathbf{l}(u) + \varepsilon \mathbf{l}^\circ(u). \tag{2.43}$$

As we know, a ruled surface can be generated by moving a line in the space. The dual version for this generation process is to move a point along the DUS. This operation requires the extension of dual matrices.

An $n \times m$ dual matrix is defined as a matrix of dual numbers in $\mathbb{D}$. We restrict to square matrices where $n = m = 3$. The operations of addition and multiplication are defined for dual matrices in the same manner as for real matrices. We restrict ourselves further to the case of dual orthogonal matrices. A dual orthogonal matrix is a matrix $\hat{O} \in \mathbb{D}^{3 \times 3}$ such that:

$$\hat{O}^T \hat{O} = \hat{O} \hat{O}^T = I, \tag{2.44}$$

where I is the identity matrix in $\mathbb{R}^{3 \times 3}$.

It can be shown [29] [59] that the displacement of lines can be represented by a dual orthogonal matrix. If a line undergoes a displacement (a rotation $[R]$ and a translation $\mathbf{d}$), the displacement operators can be represented as:

$$[\hat{R}] = [R] + \varepsilon [D][R], \tag{2.45}$$

where $[D]$ is the skew-symmetric form of $\mathbf{d}$. The displacement of the line coordinates is then written as:

$$\hat{\mathbf{X}} = [\hat{R}]\hat{\mathbf{x}}. \tag{2.46}$$

The displacement of lines can be viewed as the displacement of points on the DUS. The dual orthogonal matrices form a Lie group [53]. This property provides a computationally efficient method for generating curves on the DUS and kinematically generated ruled surfaces in $\mathbb{R}^3$.

Given two points $\hat{\mathbf{x}}_1$ and $\hat{\mathbf{x}}_2$, a curve on the DUS needs to be determined intersecting both points. A continuous motion of a line can be written as:

$$\hat{\mathbf{X}}(t) = [\hat{R}(t)]\hat{\mathbf{x}}. \tag{2.47}$$

Using the property of the Lie group[52], the first derivative of this function is:

$$\dot{\hat{\mathbf{X}}}(t) = [\dot{\hat{R}}(t)]\hat{\mathbf{x}} = [\dot{\hat{R}}(t)][\hat{R}(t)]^T \hat{\mathbf{X}} = [\hat{\Omega}(t)]\hat{\mathbf{X}}, \tag{2.48}$$

where $[\hat{\Omega}(t)]$ is a skew symmetric matrix of dual elements:

$$[\hat{\Omega}(t)] = \begin{pmatrix} 0 & -\hat{\omega}_3 & \hat{\omega}_2 \\ \hat{\omega}_3 & 0 & -\hat{\omega}_1 \\ -\hat{\omega}_2 & \hat{\omega}_1 & 0 \end{pmatrix}. \tag{2.49}$$

Given the initial condition: $\hat{\mathbf{X}}(0) = \hat{\mathbf{X}}_0$, this differential equation can be solved. the solution is:

$$\hat{\mathbf{X}}(t) = \exp([\hat{\Omega}(t)])\hat{\mathbf{X}}_0, \qquad (2.50)$$

where

$$[\hat{\Omega}(t)] = \hat{\omega} \begin{pmatrix} 0 & -\hat{g}_3 & \hat{g}_2 \\ \hat{g}_3 & 0 & -\hat{g}_1 \\ -\hat{g}_2 & \hat{g}_1 & 0 \end{pmatrix} = \hat{\omega}[ad\hat{\mathbf{g}}], \qquad (2.51)$$

where $\hat{\mathbf{g}}$ is a line, $\hat{\omega} \in \mathbb{D}$ is a dual angular velocity $\hat{\omega} = \omega + \varepsilon v$, and [ad] is an operator which takes a vector into a skew symmetric matrix. Using this notation, we can find the exponential for the Lie algebra of dual skew symmetric matrices [52]:

$$[\hat{R}(t)] = \exp(\hat{\omega}[ad\hat{\mathbf{g}}]) = [I] + \sin(\hat{\omega})[ad\hat{\mathbf{g}}] + (1 - \cos(\hat{\omega}))([ad\hat{\mathbf{g}}])^2. \qquad (2.52)$$

$[I]$ is the identity matrix in $\mathbb{R}^{3\times3}$. The angle $\hat{\omega}$ is assumed to vary linearly between the end points with respect to the independent parameter $t \in [0, 1]$. The value of the angular velocity is given by:

$$\hat{\omega} = \theta t + \varepsilon v t. \qquad (2.53)$$

When $t = 1$, $\hat{\omega}$ equals the dual angle between the points $\hat{\mathbf{x}}_1$ and $\hat{\mathbf{x}}_2$: $\hat{\mathbf{x}}_1 \cdot \hat{\mathbf{x}}_2 = \cos\hat{\omega} = x + \varepsilon x^\circ$. Then this angle is calculated by:

$$\hat{\omega} = \cos^{-1}(\hat{x}) = \cos^{-1}(x + \varepsilon x^\circ) = \cos^{-1}(x) - \varepsilon \frac{x^\circ}{\sqrt{1 - x^2}}. \qquad (2.54)$$

Then the screw axis $\hat{\mathbf{g}}$ is chosen to be perpendicular to both points $\hat{\mathbf{x}}_1$ and $\hat{\mathbf{x}}_2$:

$$\hat{\mathbf{g}} = \frac{\hat{\mathbf{x}}_1 \times \hat{\mathbf{x}}_2}{\|\hat{\mathbf{x}}_1 \times \hat{\mathbf{x}}_2\|}. \qquad (2.55)$$

Finally, the equation for the curve is given by:

$$\hat{\mathbf{x}}(t) = [\exp(ad\hat{\mathbf{g}}\hat{\omega}t)]\hat{\mathbf{x}}_1. \qquad (2.56)$$

A curve generated in this fashion is a geodesic on the DUS. Based on this process, a rule surface can be generated by the dual De Calsteljau algorithm [53] [52] [12] [13] [14] [33].

Chapter 3
Ruled Surface Approximation in Line Geometry

Suppose the given surface has already been optimized according to certain functional requirements, such as stability or efficiency, the approximated ruled surface will be a compromise between the performance and the manufacturing cost. Since we deal with the robust design, a small change in the geometric design will not affect the functional performance too much. If the error between the derived ruled surface and the original free-form surface is within a given tolerance, we can use the ruled surface instead of the original design.

In this chapter, the previous work related to the ruled surface approximation is reviewed. Those algorithms are developed based on the line geometry. We are going to give an overview of ruled surface approximation algorithms and implement one of those algorithms, which employs the characteristics of the line geometry. This algorithm is adapted and applied to approximate a given turbocharger blade surface as a ruled surface.

3.1 Approximation in Line Space

3.1.1 Algorithm overview

Ruled surface approximation can be written as a minimization problem, whose objective function is the error between the given surface and the ruled surface. In general, the objective function has the following formula:

$$F = \sum_k \|\mathbf{x}(u_k, v_k) - \mathbf{p}_k\|^2 + \lambda F_s, \tag{3.1}$$

where $\mathbf{x}(u, v)$ is a function of ruled surface and $\mathbf{p}_k$ denotes a point on the given surface. The first part is certain error measurement. F_s is a smoothing term. The crucial points of building the objective function are the choice of a ruled surface representation and an appropriate error measurement. In [6], the ruled surface is represented as a linear interpolation between two boundary curves. The distance between two lines in $\mathbb{R}^3$ is linked to the distance of two points in $\mathbb{R}^6$. The Euclidean metric in $\mathbb{R}^6$ is employed and leads to a general framework for approximation in line space. In [20], the ruled surface is represented as an envelope of a set of planes. Then an error measurement between planes is used, which leads to a linear algorithm. All the above algorithms need to find an appropriate parametrization. With inappropriate parameterizations, singularities can occur on the approximated surface in the required region of interest. In [41], an active model for the surface approximation is presented. This approach employs a second order approximation of the distance function as an error measurement, which completely avoids the parametrization problem.

3.1.2 Approximation algorithm description

In this part, we implement the algorithm introduced in [6] for the ruled surface approximation. This algorithm is used to approximate the given surface by a ruled surface in a tensor product B-spline representation, which is useful for approximating surface strips. It consists of 3 steps described below, where the third step is optional [6] [44].

The input of the algorithm is a parametric or implicit representation of a surface $\Phi \in \mathbb{R}^3$. It can be written as a bivariate function $(x, y, f(x, y))$ if there exists a vector e_3 and an angle $0 < \gamma_0 < \pi/2$, such that all surface normals form an angle $0 < \gamma < \gamma_0$. In this case, the z-axis of the Cartesian coordinate system (x, y, z) is chosen parallel to e_3, the projected domain of Φ in the $x - y$ plane is denoted as D'.

The first step of the algorithm is to find a discrete system of rulings close to the given surface. The middle point M_0 of the domain D' is chosen as a starting point. For each line segment l' passing through M_0, a corresponding curve segment c in Φ can be found by intersecting the given surface Φ with the plane strip which goes through l' and perpendicular to the $x - y$ plane. The fitted line segment l for each curve segment is constructed by minimizing the least squares error. Then the first ruling is chosen from all the fitted line segments with the smallest error, denoted as l_0. Then, for certain march direction and march distance, next middle point M_1 is found on the normal ray through M_0 of the line segment l'. By trying all lines passing through point M_1 but do not intersect the line segment l', planes passing through those lines are built perpendicular to the $x - y$ plane. Those planes will intersect the given surface with a sequence of curves. Every such curve is approximated with a line segment using the least squares fit method. The line segment with the least squares error is chosen as l_1. The march process continues till the rulings cover the whole surface domain. In the end of this step, a sequence of line segments $l_0, \ldots, l_N$ are obtained which are close to the given surface.

The second step is to construct a ruled tensor product B-spline surface approximating the ruling system. Writing a line segment by its two ending points $P = \mathbf{p} = (p_x, p_y, p_z)$ and $Q = \mathbf{q} = (q_x, q_y, q_z)$, a line segment in $\mathbb{R}^3$ is mapped to a point in $\mathbb{R}^6$:

$$\sigma : L \rightarrow \mathbb{R}^6, (\mathbf{p}, \mathbf{q}) \mapsto \mathbf{X} = (x_1, \dots, x_6) = (p_x, p_y, p_z, q_x, q_y, q_z). \quad (3.2)$$

Hence the L^2 norm of the distance function between two line segments is given by:

$$\begin{aligned} d(l_1, l_2)^2 &:= 3 \int_0^1 [(1-\lambda)(\mathbf{p_1} - \mathbf{p_2}) + \lambda(\mathbf{q_1} - \mathbf{q_2})]^2 d\lambda \\ &= [(\mathbf{p_1} - \mathbf{p_2})^2 + (\mathbf{q_1} - \mathbf{q_2})^2 + (\mathbf{p_1} - \mathbf{p_2})(\mathbf{q_1} - \mathbf{q_2})]. \end{aligned} \quad (3.3)$$

The defined distance is simply the distance of two points in $\mathbb{R}^6$ in the Euclidean metric:

$$\langle X, X \rangle = x_1^2 + \dots + x_6^2 + x_1 x_4 + x_2 x_5 + x_3 x_6. \quad (3.4)$$

Note that a B-spline curve of degree k in $\mathbb{R}^6$ corresponds to a B-spline tensor product surface of degree $(1, k)$ in $\mathbb{R}^3$. The ruled surface is represented with its boundary curves $\mathbf{p}(u)$ and $\mathbf{q}(u)$ as $(1 - v)\mathbf{p}(u) + v\mathbf{q}(u)$. It has been proven that a cubic C^2 spline is a good approximation for a ruled surface [6]. $\mathbf{p}(u)$ and $\mathbf{q}(u)$ are rewritten as a cubic B-spline in $\mathbb{R}^6$: $\mathbf{X}(u) = (\mathbf{p}(u), \mathbf{q}(u)) = \sum_{i=0}^{M} N_i^3(u)\mathbf{X}_i$. The control points $\mathbf{X}_i$ can be calculated by minimizing a function:

$$G(\mathbf{X}_0, \mathbf{X}_1, \dots, \mathbf{X}_M) := \sum_{i=0}^{N} \langle \mathbf{L}_i - \mathbf{X}(u_i), \mathbf{L}_i - \mathbf{X}(u_i) \rangle + \lambda F^*(\mathbf{X}_0, \mathbf{X}_1, \dots, \mathbf{X}_M), \quad (3.5)$$

where $F^*(\mathbf{X}_0, \dots, \mathbf{X}_M)$ is the *thin plate spline functional*:

$$F(x) = \int_a^b \int_0^1 (x_{uu}^2 + 2x_{uv}^2 + x_{vv}^2) dv du. \quad (3.6)$$

The parameters u_i influence the final result, so the chord length parametrization for the polygon $\mathbf{L}_0, \dots, \mathbf{L}_M$ in $\mathbb{R}^6$ is used as an initial guess. Then the parameters are iteratively changed using the Hoschek's method [19] until the error vectors are nearly orthogonal to the solution within the metric defined in Eq. (3.4).

After step two, an approximated ruled surface has been constructed. In the last step, improvement of the approximated ruled surface is recommended. These are a combination of least squares fits including a regularization functional such as Eq. (3.6) and parameter correction [6] [19].

In this algorithm, there are two parameters worthy to manipulate: the march distance d and the function parameter λ. The march distance controls the distance between the rulings, smaller march distance value generates more rulings for a ruled surface. It increases the accuracy for the approximation algorithm, which also increases the operation time. The function parameter λ controls the smoothness of the final boundary curve. During the current test, the λ value is chosen small, since we do not want to lose the original shape of the boundary.

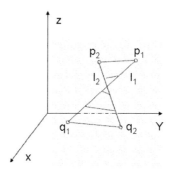

Fig. 3.1 Distance of line segments.

3.1.3 Algorithm verification

A given surface is used to test the algorithm,

$$\mathbf{S}(x,y) = (x,y,z) = (x,y,c(u) + vg(u) + r(x,y)), \tag{3.7}$$

where
$$u(x,y) = x/(6+y),$$
$$v(x,y) = y,$$
$$c(u) = 3(u^2 - 16)\sin(u),$$
$$g(u) = (4 - u^2)\cos(u),$$
$$r(x,y) = 0.5(\cos(0.3x^2)\sin(0.6xy)).$$

This surface is displayed in Fig. 3.2(a). It is not a ruled surface, since it has the extra term $r(x,y)$. This surface is used as an input to test the algorithm.

After the first step, a sequence of line segments which approximate the respective curves on the given surface have been found. These line segments are used as rulings to construct an approximated ruled surface. In Fig. 3.2(b), the blue lines are the rulings, the red points represent the ending points P, the yellow points denote the ending points Q, and the green points are the middle points of line segments. We use the coordinates for P and Q as the input for the second step.

In the second step, two cubic B-spline curves $\mathbf{p}(u)$ and $\mathbf{q}(u)$ are derived to approximate the points sequence P and Q respectively. Fig. 3.2(c) shows the two cubic B-spline curves, the blue curve is $\mathbf{p}(u)$ which approximates points P (red dots), the green curve is $\mathbf{q}(u)$ which approximates points Q (yellow dots).

Then a ruled surface is generated based on those two boundary curves $\mathbf{p}(u)$ and $\mathbf{q}(u)$ by the equation: $\mathbf{x}(u,v) = (1-v)\mathbf{p}(u) + v\mathbf{q}(u)$, $v \in [0,1]$. Fig. 3.2(d) compares

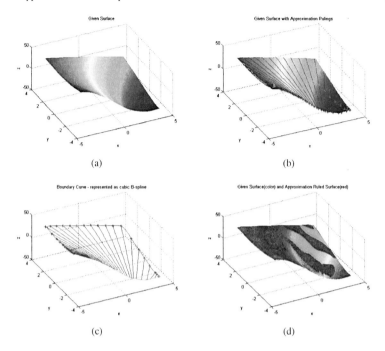

(a)   (b)

(c)   (d)

Fig. 3.2 Approximating a given surface by a ruled surface: (a) Given surface; (b) Result of step I: discrete ruling system; (c) Approximated ruled surface; (d) Comparison between the given surface and the ruled surface.

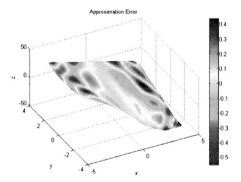

Fig. 3.3 Error between the ruled surface and the given surface.

the generated ruled surface with the given surface. Here, the surface displayed in red color is the approximated ruled surface, which is derived from the given surface.

Fig. 3.3 shows the approximation error between the ruled surface and the given surface. From the test results, we can declare that the approximation error of this algorithm is acceptable.

3.2 Application in Turbocharger Blade Design

In the previous section, an approximation algorithm based on line geometry is implemented and tested with a given surface. This algorithm can be adapted and applied to approximate a turbocharger blade as a ruled surface. In this section, some background knowledge of this application is introduced, then the blade manufacturing method is explained, finally a real case is adopted for simulation purpose.

3.2.1 Background knowledge

A turbocharger is usually used in an automobile to increase the engine power. A typical turbocharger consists of a turbine and a compressor linked by a shared axle. The turbine inlet receives exhaust gases from the engine causing the turbine wheel to rotate. This rotation drives the compressor, compressing air and delivering it to the inter-cooler which cools the air, then the colder air with higher pressure is delivered into the cylinder. Because the turbocharger increases the pressure at the point where air is entering the cylinder, more air will be forced in as the inlet manifold pressure increases. The additional air makes it possible to add more fuel, increasing the power and torque output of the engine, particularly at high engine rotation speeds [61].

A turbocharger has four main components: the turbine and impeller/compressor wheels, the impeller/compressor housings, the Split-inlet exhaust housings, and the center hub rotating assembly houses. Fig. 3.4[1] shows one example of turbochargers [21]. Here, we focus on the turbine and impeller/compressor wheels design. The turbine and impeller wheel sizes dictate the amount of air or exhaust that can be flowed through the system, and the relative efficiency at which they operate. Generally, the larger the turbine wheel and impeller wheel, the larger the flow capacity. Measurements and shapes can vary, as well as the curvature and the number of blades on the wheels.

Impellers can be classified into two types: splitter type and non-splitter type. The shape of the splitter type impeller is composed of pressure and suction surfaces, splitter, trailing edge, and leading edge as shown in Fig. 3.5 [21]. If the impeller contains no splitter, it is a non-splitter type. Impellers are modeled using section curves, just like Fig. 3.6 [21] shows. Considering the fluid flow and manufacturing cost, blade and splitter surfaces are usually designed in ruled surfaces, which can be manufactured using the flank (side) milling method. This method uses the side of a

---

[1] Taken from Wikipedia.

Fig. 3.4 An example of turbocharger.

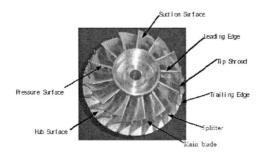

Fig. 3.5 Components of an impeller.

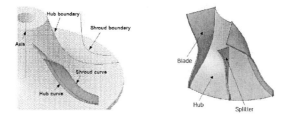

Fig. 3.6 Model of an impeller.

manufacturing tool to remove the material, which does not leave any cutting marks. The detail of the flank milling method is introduced in Chapter 8. Flank milling tool paths of blades and splitters are shown in Fig. 3.7 [21].

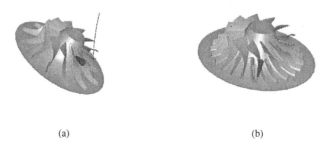

(a)   (b)

Fig. 3.7 Ruled surface manufacturing: (a) Manufacture blade of ruled surface; (b) Manufacture splitter of ruled surface.

3.2.2 Simulation result

We modified the algorithm to approximate a given turbocharger blade surface with a ruled surface. The main modification is about the first step in order to fit the input. The input file is a "*.ibl" file from the software "Bladegen". It contains discrete points on different blade layers. Those points are first fitted to a surface by applying certain interpolation method. Then the algorithm introduced in the previous section can be applied to construct the approximated ruled surface. Fig. 3.8(a) shows one side surface of a given blade. Fig. 3.8(b) displays the approximated rulings of the given surface. Fig.3.8(c) draws two approximated boundary curves for the end points of rulings. Fig. 3.8(d) compares the approximated ruled surface with the given surface. Fig. 3.9 shows the error between the obtained ruled surface and the given blade surface. Through calculation, the maximum error in z direction is 0.1192mm and the average error is 0.0062mm, the ratio between the average error and the surface scale is very small (4.3446e-004). Therefore, the obtained ruled surface can be used instead of the original design.

3.3 Comments

In this chapter, we reviewed the previous work of ruled surface approximation algorithms and implemented one typical approximation algorithm in line geometry. The algorithm is verified using a given surface. This algorithm can be modified for certain industrial application. The background of turbocharger and blade design is presented.

Using this algorithm, an approximated ruled surface is derived. It is then delivered to the manufacturing phase. The manufacturing method leads automatically to the requirement of kinematic ruled surface approximation algorithms. Planning the

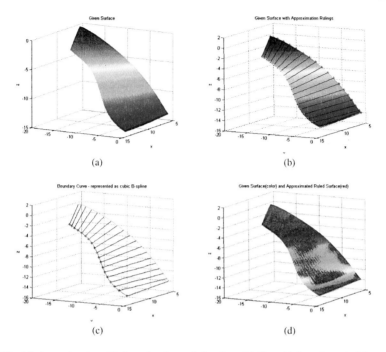

(a)   (b)

(c)   (d)

Fig. 3.8 Approximating a blade surface by a ruled surface: (a) Given blade surface; (b) Discrete ruling system; (c) Approximated boundary curves; (d) Comparison between the given blade surface and the approximated ruled surface.

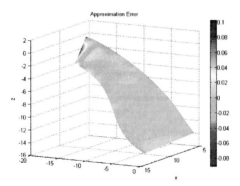

Fig. 3.9 Error between the ruled surface and the given blade surface.

tool motion of flank milling for a ruled surfaces is a kinematic approximation problem. The goal is to develop a method for synthesizing a motion such that the swept surface of such a motion with a cylindrical cutter closely approximates a given ruled surface. Generally, a motion is described by a Bézier curve or a B-spline, which is used to interpolate or approximate a set of discrete cutter locations for the flank milling of a given ruled surface. Then the control positions of the motion is fine-tuned to minimize the error between the swept surface of the cylindrical cutter under the motion and the desired ruled surface. A kinematic ruled surface approximation algorithms provide a direct guide of the milling tool movement.

Chapter 4
Theory of the Kinematic Ruled Surface Approximation

Based on the knowledge in Chapter 2, the coordinates of lines can be written in dual vectors through the Klein mapping and the Study mapping. In this manner, a ruled surface is a one-parameter curve on the dual unit sphere. The problem of approximating/interpolating ruled surfaces is reduced to the problem of approximating/interpolating curves on the dual unit sphere.

4.1 Ruled Surface Mapping Approach

In Chapter 2, we derive the dual vector representation for a line in P^3 space. In screw and dual number algebra, the Study displace theory shows that there is a one-to-one correspondence between an oriented straight line in $\mathbb{R}^3$ and a point on the DUS [64].

4.1.1 Dual vector representation of ruled surface

As we know, an oriented line $\mathbf{L}$ in $\mathbb{R}^3$ can be represented by a Plücker vector coordinate $(\mathbf{l}, \mathbf{l}^\circ)$. The vector $\mathbf{l}$ indicates the direction of $\mathbf{L}$, which is usually normalized to $\|\mathbf{l}\| = 1$. The vector $\mathbf{l}^\circ$ is the moment vector with respect to the origin. Hence, a line $\mathbf{L}$ can be expressed by a dual vector:

$$\hat{l} = \mathbf{l} + \varepsilon \cdot \mathbf{l}^\circ = \mathbf{l} + \varepsilon \cdot \mathbf{OP} \times \mathbf{l} = \mathbf{l} + \varepsilon \cdot \mathbf{OD} \times \mathbf{l}, \tag{4.1}$$

where $\mathbf{OP}$ is a point vector on the line $\mathbf{L}$ and $\mathbf{OD} \perp \mathbf{L}$.

A ruled surface can be viewed as a motion locus of a straight line, therefore it can be described by a dual curve which is generated by the locus of the corresponding dual points on the DUS:

$$\mathbf{L}(u) = \hat{\mathbf{x}}(u) = \mathbf{l}(u) + \varepsilon \mathbf{l}^\circ(u). \tag{4.2}$$

4.1.2 Dual vector ruled surface representation transform

For a given dual vector representation of a ruled surface, a representation in $\mathbb{R}^3$ can be derived. Eq. (4.2) is rewritten as:

$$\hat{\mathbf{x}}(u) = \mathbf{l}(u) + \varepsilon \mathbf{l}^\circ = \mathbf{l}(u) + \varepsilon \mathbf{OP}(\mathbf{u}) \times \mathbf{l} = \mathbf{l}(u) + \varepsilon \mathbf{OD}(\mathbf{u}) \times \mathbf{l}. \tag{4.3}$$

Because $\mathbf{OD}(\mathbf{u}) \perp \mathbf{L}(u)$ and $\mathbf{D}$ is a point on line $\mathbf{L}(u)$, we get the following equations:

$$l_2(u)x_{3D}(u) - l_3(u)x_{2D}(u) = l_1^\circ(u), \tag{4.4a}$$

$$l_3(u)x_{1D}(u) - l_1(u)x_{3D}(u) = l_2^\circ(u), \tag{4.4b}$$

$$l_1(u)x_{2D}(u) - l_2(u)x_{1D}(u) = l_3^\circ(u), \tag{4.4c}$$

$$l_1(u)x_{1D}(u) + l_2(u)x_{2D}(u) + l_3(u)x_{3D}(u) = 0, \tag{4.4d}$$

$(\mathbf{l}, \mathbf{l}^\circ) = (l_1(u), l_2(u), l_3(u), l_1^\circ(u), l_2^\circ(u), l_3^\circ(u))$, $\mathbf{OD}(u) = (x_{1D}(u), x_{2D}(u), x_{3D}(u))$. Solving the above system of Eq. (4.4), the coordinate of $\mathbf{OD}(u)$ is derived:

$$\begin{aligned}
x_{1D}(u) &= l_2(u)l_3^\circ(u) - l_3(u)l_2^\circ(u), \\
x_{2D}(u) &= l_3(u)l_1^\circ(u) - l_1(u)l_3^\circ(u), \\
x_{3D}(u) &= l_1(u)l_2^\circ(u) - l_2(u)l_1^\circ(u).
\end{aligned} \tag{4.5}$$

The parametric equation of a ruled surface in $\mathbb{R}^3$ can be expressed as:

$$\mathbf{x}(u, v) = \mathbf{r}_D(u) + v\mathbf{l}(u) = \mathbf{l}(u) \times \mathbf{l}(u)^\circ + v\mathbf{l}(u), \tag{4.6}$$

where $\mathbf{r}_D(u) = (x_{1D}(u), x_{2D}(u), x_{3D}(u))$. Taking two parameters δ_{p1}, δ_{p2} of a straight line, two directrix curves of a ruled surface can be written as:

$$\mathbf{r}_{p1}(u) = \mathbf{l}(u) \times \mathbf{l}(u)^\circ + \delta_{p_1}(u)\mathbf{l}(u), \tag{4.7a}$$

$$\mathbf{r}_{p2}(u) = \mathbf{l}(u) \times \mathbf{l}(u)^\circ + \delta_{p_2}(u)\mathbf{l}(u). \tag{4.7b}$$

4.2 Weighted Average on a Real Sphere and Real Spherical Spline

In the previous section, we established a transformation mapping between a ruled surface representation in Euclidean space and a curve representation on the DUS. Instead of solving a surface approximation/interpolation problem in Euclidean space, we solve a curve approximation/interpolation problem in the DUS space.

In order to find an algorithm for curve approximation/interpolation on the DUS, we first studied the algorithms for curve approximation/interpolation on the real sphere. There are several types of algorithms. One typical algorithm is adding a constrain to the objective function which makes sure that the fitted curve lying on the sphere [17]. Besides the constrained optimization algorithm, another idea is to employ spherical polynomials as the bases for curve fitting [63] [51]. Recently, a spherical spline space is constructed by defining Bernstein-Bézier polynomials on triangulations on the sphere, and planar interpolation algorithms can be applied on sphere-like surfaces [2] [1]. Even further, certain metric on the sphere-like surface is constructed, which can be used to build an objective function for curve fitting. However, those algorithms are either too slow or too complicated, we are searching an algorithm which is fast and easy to extend in the dual number space, besides, standard Bernstein-Bézier polynomials are preferred, since they are widely used in industrial design and manufacturing.

The idea in this chapter is inspired by the paper [5], in which a method for computing weighted averages on spheres based on the least squares minimization is proposed. The weighted average method allows a novel method of defining Bézier and spline curves on spheres, which provides a direct generalization of Bézier and B-spline curves to spherical spline curves. Based on this idea, a fast algorithm for the spline interpolation on spheres is developed. Before we introduce our definition of dual spherical spline, we briefly describe the essence of the paper [5].

4.2.1 Spherical weighted average

Let $\mathbf{p}_1,\ldots,\mathbf{p}_n$ be points on a d-dimensional unit sphere S^d in $\mathbb{R}^{d+1}$: a weighted average of these n points using weight values $\omega_1,\ldots,\omega_n$ such that each $\omega_i \geq 0$ and $\sum \omega_i = 1$ is denoted as:

$$\mathbf{C} = \widetilde{\sum}_{i=0}^{n} \omega_i \mathbf{p}_i. \tag{4.8}$$

This weighted average is to be defined in terms of distances in S^d. Since S^d is not a linear space, the weighted average in Eq. (4.8) is not simply a linear combination of the points $\mathbf{p}_1,\ldots,\mathbf{p}_n$. It is defined as a result of least squares minimization, namely as the point $\mathbf{C}$ on S^d which minimizes the value:

$$f(\mathbf{C}) = \frac{1}{2} \sum_i \omega_i \cdot dist_S(\mathbf{C},\mathbf{p}_i)^2. \tag{4.9}$$

where $dist_S(\mathbf{C},\mathbf{p}_i)$ is the spherical distance between $\mathbf{C}$ and $\mathbf{p}_i$. The function f attains a unique minimum if the following condition satisfied.

theorem 4.2.1 *Suppose the points $\mathbf{p}_1,\ldots,\mathbf{p}_n$ all lie in a hemisphere H of S^d, with at least one point $\mathbf{p}_i$ in the interior of H with $\omega_i \neq 0$. Then the function f has a single critical point $\mathbf{q}$ in H, and this point $\mathbf{q}$ is the global minimum of f.*

The assumption that the points all lie on a single hemisphere can be relaxed significantly. In fact, it only requires that the second derivative of the objective function is positive. In precise, the uniqueness requirement is described as follows, for the proof details please refer to [5].

theorem 4.2.2 *Let f be the objective function, $\mathbf{p}_i$ are the given points on the unit sphere and ω_i are the weights, $B_q(t)$ denotes the ball of radius t around q.*

(a) Let $0 < \phi < \pi/2$. Suppose $\mathbf{q}$ is a critical point of f and $B_q(\phi)$ contains points $\mathbf{p}_i$ of total weight ≥ 0.5 and then all the points $\mathbf{p}_i$ are contained in the $B_q(\pi - \phi)$. Then $\mathbf{q}$ is a local minimum of f according to the second derivative test.

(b) Let $0 < \phi < \psi < \pi/2$. Suppose $0 < \omega \leq 1$ and that $\omega\phi + (1 - \psi)\pi \leq \psi$. Further suppose there is a point $\mathbf{v}$ so that $B_v(\phi)$ contains points $\mathbf{p}_i$ of total weight $\geq \omega$ and that all the points $\mathbf{p}_i$ are in $B_v(\pi - 2\psi - \phi)$. Then f has a unique minimum. This unique minimum is inside $B_v(\psi)$ and is the only critical point inside $B_v(\psi)$.

4.2.2 Properties of the spherical weighted average

After deriving the definition of a spherical weighted average, the uniqueness of the definition is checked. It can be proven that the new defined spherical weighted average has nice properties. Here, we mention two theorems: one concerns the continuity of spherical weighted average and the other shows the convexity property of the spherical weighted average [5]. First, we describe the continuity theorem, which can be proven by the Implicit Function Theorem:

theorem 4.2.3 *Let values for $\mathbf{p}_1, \ldots, \mathbf{p}_n$ and $\omega_1, \ldots, \omega_n$ and $\mathbf{q}$ be chosen that satisfy the hypotheses of Theorem 4.2.1 or 4.2.2. Then there is a neighborhood of $\mathbf{p}_1, \ldots, \mathbf{p}_n, \omega_1, \ldots, \omega_n$ in which the weighted average $\mathbf{q}$ is a C^∞-function of $\mathbf{p}_1, \ldots, \mathbf{p}_n, \omega_1, \ldots, \omega_n$.*

Before discussing the convexity property of the spherical weighted average, the definition of the *convex set* and the *convex hull* should be set up.

definition 4.2.1 *A subset C of S^d is convex iff for any two points $x, y \in C$, there is a shortest geodesic from x to y which lies entirely in C. The subset C is the convex hull of a set D iff C is the unique smallest convex set containing D.*

It can be shown that the points $\mathbf{q}$ which can be written as a weighted average of $\mathbf{p}_1, \ldots, \mathbf{p}_k$ form a convex set, in fact, they form precisely the convex hull of points $\mathbf{p}_1, \ldots, \mathbf{p}_k$. $\mathbf{q}$ is a proper weighted average of $\mathbf{p}_1, \ldots, \mathbf{p}_k$ if there are non-negative weights $\omega_1, \ldots, \omega_k$ with sum to 1 and for each non-zero ω_i, $\mathbf{p}_i$ is interior of a hemisphere $\mathcal{H}$. The following theorem holds:

theorem 4.2.4 *Suppose that $\mathbf{p}_1, \ldots, \mathbf{p}_k$ are distinct points, and that not the case that $k = 2$ with $\mathbf{p}_1$ and $\mathbf{p}_2$ antipodal. Then the convex hull C of $\{\mathbf{p}_1, \ldots, \mathbf{p}_k\}$ exists and is equal to the set of proper weighted averages of $\mathbf{p}_1, \ldots, \mathbf{p}_k$. If $\mathbf{p}_1, \ldots, \mathbf{p}_k$ lie in a hemisphere, then the convex hull C is a subset of the hemisphere. If they do not lie in a hemisphere, then the convex hull is the entire sphere S^d.*

Following the proof of the previous theorem, two more theorems are derived as a corollary.

theorem 4.2.5 *Suppose that* $\mathbf{p}_1, \ldots, \mathbf{p}_k$ *are distinct points, and that it not the case that $k = 2$ with $\mathbf{p}_1$ and $\mathbf{p}_2$ antipodal. Then the convex hull C of $\{\mathbf{p}_1, \ldots, \mathbf{p}_k\}$ exists and is equal to the set of strongly proper weighted averages of $\mathbf{p}_1, \ldots, \mathbf{p}_k$.*

theorem 4.2.6 *Every point in the convex hull C of $\{\mathbf{p}_1, \ldots, \mathbf{p}_k\}$ can be written as a strongly proper weighted average of at most $d + 1$ many of $\mathbf{p}_1, \ldots, \mathbf{p}_k$.*

4.2.3 Spherical spline

Based on the definition of a weighted average on the real unit sphere, the spline functions which take values on the unit d-sphere S^d can be defined analogously. The most common method of using splines in Euclidean space $\mathbb{R}^{d+1}$ involves the selection of control points $\mathbf{p}_1, \ldots, \mathbf{p}_n$ in $\mathbb{R}^{d+1}$ and basis functions $f_1(t), \ldots, f_n(t)$. The Euclidean spline curve $\mathbf{s}(t)$ is defined by

$$\mathbf{s}(t) = \sum_{i=1}^{n} f_i(t)\mathbf{p}_i. \tag{4.10}$$

The basis functions must always satisfy the property:

$$\sum_{i=1}^{n} f_i(t) = 1, \ f_i(t) \geq 0 \ \forall i, \tag{4.11}$$

for t in the interval $[a, b]$. Usually the basis functions have additional properties such as having continuous k-th order derivatives or having a local support.

In order to define spherical splines, let $\mathbf{p}_1, \ldots, \mathbf{p}_n$ now be points on S^d, and let $f_1(t), \ldots, f_n(t)$ be functions which satisfy property 4.11. the spline curve $\mathbf{s}(t)$ which takes values on the unit sphere is define as:

$$\mathbf{s}(t) = \widetilde{\sum}_{i=1}^{n} f_i(t)\mathbf{p}_i. \tag{4.12}$$

In practice, for each value of the parameter t, the set of control points $\mathbf{p}_i$ for which $f_i(t) \neq 0$ is contained inside a hemisphere, or at least is mostly contained inside a hemisphere so as to satisfy the conditions of uniqueness. This condition is most readily met provided that either (a) the basis functions have local support and consecutive control points are not too widely spaced, or (b) the control points all lie inside a single hemisphere.

The most common applications of splines use B-splines with the basis functions f_i, which are piecewise cubic curves with continuous second derivatives. From Theorem 4.2.3, we know that if the basis functions f_i have continuous k-th derivatives, then the spline curve also has k-th derivatives. In this case the spherical spline points $\mathbf{s}(t)$ will be well-defined provided that any four consecutive control points lie in a hemisphere.

4.3 Weighted Average on the Dual Unit Sphere and Dual Spherical Spline

An oriented line in Euclidean space can be represented as a dual unit vector, which is a point on the DUS in $\mathbb{D}^3$. Writing the Plücker coordinate of a line in the projective space as a dual unit vector, the computational problem in a quadric of P^5 is reduced to a problem in a dual form of spherical geometry. In this section, we define the spherical average on the DUS and the dual spherical splines. The uniqueness of the definition is verified. The continuity and convexity properties are discussed based on the transfer principle [11].

4.3.1 Definition of the weighted average on the DUS

Based on the transfer principle of dual unit vectors, which simply states that for any operation defined for a real vector space, there is a dual version with similar interpretation, we can derive a similar definition of a spherical average on the DUS. Since we are only interested in the case for the DUS in $\mathbb{D}^3$, the definition is restricted as follows:

definition 4.3.1 *Let $\hat{\mathbf{p}}_1, \ldots, \hat{\mathbf{p}}_n$ be points on the dual unit sphere $\hat{S}^2$ in $\mathbb{D}^3$: a weighted average of these n points using real weight values $\omega_1, \ldots, \omega_n$, such that each $\omega_i \geq 0$ and $\sum_i \omega_i = 1$, is defined as the point $\hat{\mathbf{q}}$ on $\hat{S}^2$ which minimizes the value:*

$$\hat{f}(\hat{\mathbf{q}}) = \frac{1}{2}\sum_i \omega_i \cdot dist_{\hat{S}}(\hat{\mathbf{q}}, \hat{\mathbf{p}}_i)^2, \tag{4.13}$$

where $dist_{\hat{S}}(\hat{\mathbf{q}}, \hat{\mathbf{p}}_i)$ is the dual spherical distance between $\hat{\mathbf{q}}$ and $\hat{\mathbf{p}}_i$. $\hat{f}$ is a dual number function, which follows the lexicographical order:

$$a + \varepsilon a^\circ < b + \varepsilon b^\circ \text{ if and only if } a < b \text{ or } a = b \text{ and } a^\circ < b^\circ. \tag{4.14}$$

The weighted average on the DUS is denoted as:

$$\hat{\mathbf{q}} = \widetilde{\sum}_{i=0}^n \omega_i \hat{\mathbf{p}}_i. \tag{4.15}$$

The distance between two points on the DUS is defined by a dual angle between two lines. It has the form $\hat{\theta} = \theta + \varepsilon d$, where θ is the angle between the lines and d is the minimum distance along the common perpendicular. So for two points $\hat{\mathbf{x}}$ and $\hat{\mathbf{y}}$ on the DUS, we have the following equation:

$$\hat{\mathbf{x}} \cdot \hat{\mathbf{y}} := x + \varepsilon x^\circ = \cos \hat{\theta}. \tag{4.16}$$

In Chapter 2, we have known that a differentiable function $f(x)$ can be defined for a dual argument $f(x + \varepsilon x^\circ)$ by expanding the function using Taylor series. Since

$\varepsilon^2 = 0$, the dual arc-cosine function is defined as:

$$\hat{\theta} = \cos^{-1}(x + \varepsilon x^\circ) = \cos^{-1}(x) - \varepsilon \frac{x^\circ}{\sqrt{1 - x^2}}. \tag{4.17}$$

4.3.2 Existence and uniqueness of the weighted average on the DUS

The existence and uniqueness theorem of the definition 4.13 is stated similarly:

theorem 4.3.1 *Suppose the points $\hat{\mathbf{p}}_1, \ldots, \hat{\mathbf{p}}_n$ all lie on a dual hemisphere $\hat{H}$ of $\hat{S}^2$, i.e., the spherical distance $\hat{\theta} := \theta + \varepsilon d$ between arbitrary two points $\hat{\mathbf{p}}_i$ and $\hat{\mathbf{p}}_j$ satisfies $0 \leq \theta \leq \pi$, with at least one point $\hat{\mathbf{p}}_i$ in the interior of $\hat{H}$ with $\omega_i \neq 0$. Then the function $\hat{f}$ has a single critical point $\hat{\mathbf{q}}$ in $\hat{H}$, and this point $\hat{\mathbf{q}}$ is the global minimum of $\hat{f}$.*

To prove the theorem 4.3.1, the same strategy as [5] is adopted. Before the proof, the exponential and logarithmic maps are defined for dual vectors. These maps are useful for the proof as well as the algorithm development.

4.3.2.1 Exponential and logarithmic maps

Firstly, we define a subspace of $\mathbb{D}^3$, represented as:

$$T := \{\hat{\mathbf{x}} \mid \hat{\mathbf{x}} = (\hat{x}_1, \hat{x}_2, 0); \hat{x}_1, \hat{x}_2 \in \mathbb{D}\}. \tag{4.18}$$

Obviously, the subspace T is a linear space. Eq. (2.38) is still valid in the subspace. This subspace can be considered as a tangent hyperplane with respect to a point $\hat{\mathbf{q}}$ on the DUS. Without loss of generality, we choose the point as $\hat{\mathbf{q}} := (0, 0, 1)$, and the points on the tangent plane of point $\hat{\mathbf{q}}$ can be written as $\hat{\mathbf{x}} := (\hat{x}_1, \hat{x}_2, 1)$. The subspace T is derived by setting the point $\hat{\mathbf{q}}$ as the origin of T_q. The distance between $\hat{\mathbf{q}}$ and $\hat{\mathbf{p}}_i$ on the hyperplane is calculated as:

$$\hat{r} = \|\hat{\mathbf{x}} - \hat{\mathbf{q}}\| = \|(\hat{x}_1, \hat{x}_2, 0)\|. \tag{4.19}$$

The exponential map at $\hat{\mathbf{q}}$ is defined to map points from the tangent hyperplane T_q to the DUS, which preserves angles and distances from $\hat{\mathbf{q}}$. The exponential map is denoted as $\exp_{\hat{\mathbf{q}}}(\cdot)$. For this case, it is a function mapping a point $\hat{\mathbf{p}}$ with coordinate $(\hat{x}_1, \hat{x}_2, 1)$ to a point $\exp_{\hat{\mathbf{q}}}(\hat{\mathbf{p}}) = (\hat{x}_1', \hat{x}_2', \hat{x}_3')$.

The following condition should be satisfied to preserve the distance:

$$\hat{x}_3' = \cos(\hat{r}), \tag{4.20}$$

where $\hat{r}$ is define by Eq. (4.19). Since $\exp_{\hat{q}}(\hat{p})$ is located on the DUS, applying the property $\sin^2(\hat{r}) + \cos^2(\hat{r}) = 1$, we define:

$$\hat{x}_1' = \hat{x}_1 \cdot \frac{\sin(\hat{r})}{\hat{r}} \text{ and } \hat{x}_2' = \hat{x}_2 \cdot \frac{\sin(\hat{r})}{\hat{r}}. \tag{4.21}$$

In case $r = 0$, the dual quotient is not defined, we assign $\hat{x}_1' = \hat{x}_1$ and $\hat{x}_2' = \hat{x}_2$.

The logarithmic map is the inverse of the exponential map, which maps a point $\hat{p}' = (\hat{x}_1', \hat{x}_2', \hat{x}_3')$ on the DUS to a point $(\hat{x}_1, \hat{x}_2, 1)$ on the tangent hyperplane $T_{\hat{q}}$, provided $\hat{p}'$ and $\hat{q}$ are on the same hemisphere. The logarithmic map is denoted as $l_{\hat{q}}(\cdot)$. Based on the relation $\exp_{\hat{q}}(l_{\hat{q}}(\hat{p}')) = \hat{p}'$, the inverse map is defined as:

$$\hat{x}_i = \hat{x}_i' \cdot \frac{\hat{\theta}}{\sin(\hat{\theta})}; \text{ for } i = 1, 2, \tag{4.22}$$

where $\hat{\theta} = \cos^{-1}(\hat{x}_3')$ is the dual angle between $\hat{p}'$ and $\hat{q}$ ($\hat{p}'$ and $\hat{q}$ are on the same hemisphere, i.e., the principal part θ of the dual angle $\hat{\theta}$ satisfies $0 \leq \theta \leq \pi$). In case $\sin(\theta) = 0$, $\hat{x}_i = \hat{x}_i'$ for $i = 1, 2$.

4.3.2.2 Proof of existence and uniqueness

The dual unit sphere is an image of the Klein quadric M_2^4 [44], and the image of any quadric is compact [4]. According to the theorem of *continuous image of a compact set*: If K is compact and f is a continuous functions defined on K, then $f(K)$ is compact [56]. Since $\hat{f}$ is a continuous function in the compact metric space $\hat{S}^2$, which is complete and totally bounded, $\hat{f}$ attains its minimum value at least at one point $\hat{q}$. It can be proven that $\hat{q}$ is in the interior of a hemisphere $\hat{H}$.

Suppose that $\hat{q}$ is a solution of the minimum of Eq. (4.13) and lies completely outside $\hat{H}$, we can find a point $\hat{q}'$ in the interior of $\hat{H}$ by reflecting $\hat{q}$ across the boundary of $\hat{H}$ (antipodal point). The antipodal points on the DUS represent the same line in $\mathbb{R}^3$. For an infinite line, the orientation vector is unrestricted. Opposite orientations of this vector correspond to the antipodal points on the DUS. For point $\hat{p}_i$ inside the hemisphere, the distance between $\hat{p}_i$ and $\hat{q}'$ is denoted as a dual angle $\theta + \varepsilon d$. Then the distance between $\hat{p}_i$ and $\hat{q}$ is $(\theta + \pi) + \varepsilon d$, so $\hat{f}(\hat{q}') < \hat{f}(\hat{q})$. It contradicts the assumption. Therefore, $\hat{q}$ can not line outside $\hat{H}$.

Next, we prove $\hat{q}$ can not lie on the boundary of $\hat{H}$ either. It is equivalent to prove that the gradient of $\hat{f}$ at the boundary is always non-zero and is pointing outwards from $\hat{H}$. From chapter 2, we know that the dual vector representation of line is a screw. To analyze the gradient of the function $\hat{f}$, we flow the rules in screw calculus [11].

$\hat{f}$ is a function of a screw (or dual vector) $\hat{s}$, which has the form:

$$\hat{f}(\hat{s}) = \hat{f}(s + \varepsilon s^\circ). \tag{4.23}$$

We express the argument in terms of dual vectors in a rectangular coordinate system with origin as point O and then apply the formulas for a dual number argument. The dual coordinate elements are:

$$\hat{s}_x = s_x + \varepsilon s_x^\circ \,,\; \hat{s}_y = s_y + \varepsilon s_y^\circ \,,\; \hat{s}_z = s_z + \varepsilon s_z^\circ \,, \tag{4.24}$$

where $s_x, s_y, s_z, s_x^\circ, s_y^\circ, s_z^\circ$ are six real elements of a Plücker coordinate vector. Expanding the function using Taylor series and adopting the differentiation rules for the dual number function, we get:

$$\begin{aligned}
\hat{f}(\hat{s}_x, \hat{s}_y, \hat{s}_z) &= \hat{f}(s_x + \varepsilon s_x^\circ, s_y + \varepsilon s_y^\circ, s_z + \varepsilon s_z^\circ) \\
&= \hat{f}(s_x, s_y, s_z) + \varepsilon \Big(s_x^\circ \frac{\partial \hat{f}}{\partial s_x} + s_y^\circ \frac{\partial \hat{f}}{\partial s_y} + s_z^\circ \frac{\partial \hat{f}}{\partial s_z} \Big).
\end{aligned} \tag{4.25}$$

The function $\hat{f}$ becomes real if all variables are real, then $\hat{f}(s_x, s_y, s_z) = f(s_x, s_y, s_z)$. Returning to the vector notation, we get:

$$\hat{f}(\hat{s}) = f(s) + \varepsilon s^\circ \cdot \nabla f(s) = f(s) + \varepsilon (s^\circ \cdot \nabla) f(s). \tag{4.26}$$

Analyzing the previous equation, we note that the function $\hat{f}(\hat{s})$ is fully defined by a function of its principal part $f(s)$. Hence, if two dual vector functions $F(\hat{x})$ and $\Phi(\hat{x})$ satisfy $\nabla F(x) = \Phi(x)$, then $\nabla F(\hat{x}) = \Phi(\hat{x})$ holds [11].

To prove that the gradient of $\hat{f}$ at the boundary is always non-zero and is pointing outwards from $\hat{H}$, it is equivalent to prove that the gradient of the real vector function $f(x)$ at the boundary is always non-zero and is pointing outwards from the real hemisphere H, which has been done in [5]. Till here, we prove that $\hat{f}$ attains its minimum value at least at one point $\hat{q}$ and $\hat{q}$ is in the interior of the hemisphere $\hat{H}$.

More generally, we quote the following theorem [11]:

theorem 4.3.2 *All formulas and all theorems of vector analysis remain valid in the domain of screws.*

It follows from all the above that a dual vector function analysis can be constructed by substitution of dual vectors for vectors. The correspondence between geometrical objects is preserved: the dual modulus of the screw corresponds to the modulus of the vector and the dual angle between the axes of the screws corresponds to the angle between vectors.

For the uniqueness proof, our task is to verify that the second derivative of $\hat{f}$ at point $\hat{q}$ is positive. Its second-order derivatives at $\hat{q}$ are equal to $\left(\frac{\partial^2 \hat{F}}{\partial \hat{x}_i \partial \hat{x}_j}\right)_{\hat{q}}$. According to theorem 4.3.2, instead of calculating $\left(\frac{\partial^2 \hat{F}}{\partial \hat{x}_i \partial \hat{x}_j}\right)_{\hat{q}}$, we prove that the second derivative of the principal part $f(x)$ is positive. It follows exactly the same procedure as the uniqueness proof of a weighted average on a real sphere. The details of the proof can be found in [5].

4.3.3 Properties of the weighted average on the DUS

Using the transfer principle in the screw theory, we can get the continuity property of the dual spherical average. It has been proven that the differentiation property of a screw function is fully decided by its principle part. Hence, we have the similar continuity theorem as the real sphere case:

theorem 4.3.3 *Let values for* $\hat{\mathbf{p}}_1, \ldots, \hat{\mathbf{p}}_n$ *and* $\omega_1, \ldots, \omega_n$ *and* $\hat{\mathbf{q}}$ *be chosen that satisfy the hypotheses of Theorem 4.3.1. Then there is a neighborhood of* $\hat{\mathbf{p}}_1, \ldots, \hat{\mathbf{p}}_n$, $\omega_1, \ldots, \omega_n$ *in which the weighted average* $\hat{\mathbf{q}}$ *is a* C^∞*-function of* $\hat{\mathbf{p}}_1, \ldots, \hat{\mathbf{p}}_n, \omega_1, \ldots, \omega_n$.

It can also be shown that the points $\hat{\mathbf{q}}$ which can be written as a weighted average of $\hat{\mathbf{p}}_1, \ldots, \hat{\mathbf{p}}_k$ form a convex set, they form precisely the convex hull of points $\hat{\mathbf{p}}_1, \ldots, \hat{\mathbf{p}}_k$.

4.3.4 Definition of the dual spherical spline

Based on the definition of a weighted average on the DUS in $\mathbb{D}^3$, the spline functions which take values on the Dual Unit Sphere can be defined analogously. Same as the definition of spherical splines on the real sphere, the basis functions must always satisfy the property:

$$\sum_{i=1}^{n} f_i(t) = 1, \; f_i(t) \geq 0 \; \forall i, \tag{4.27}$$

for t in the interval $[a, b]$. However $\hat{\mathbf{p}}_1, \ldots, \hat{\mathbf{p}}_n$ are now points on the dual unit sphere in $\mathbb{D}^3$, the spline curve $\hat{\mathbf{s}}(t)$ which takes values on the dual unit sphere is defined as:

$$\hat{\mathbf{s}}(t) = \widetilde{\sum}_{i=1}^{n} f_i(t)\hat{\mathbf{p}}_i. \tag{4.28}$$

In order to satisfy the uniqueness requirement, for each value of the parameter t, the set of control points $\hat{\mathbf{p}}_i$ for which $f_i(t) \neq 0$ is contained inside a dual hemisphere.

4.4 Comments

In this chapter, we first introduce the definition of a weighted average on the real sphere and its application on spherical splines. Using the screw theory and transfer principal, we define a weighted average on the dual unit sphere. The uniqueness of the definition is verified. The continuity property and convexity property are preserved on the DUS. Therefore, it leads automatically to the definition of dual spherical splines. These definitions lay the basis for dual spherical spline interpolation algorithms, which will be introduced in the next chapter.

Chapter 5
Dual Spherical Spline Interpolation Algorithm

In Chapter 4, a weighted average on a real sphere is introduced and a weighted average on the dual unit sphere is defined. This new definition leads to a well defined dual spherical spline. In this chapter, we first implement the algorithm calculating the weighted average on the real sphere. It is extended to the algorithm interpolating the spherical spline on the real sphere. Then, a new algorithm interpolating dual spherical splines on the dual unit sphere is proposed.

5.1 Algorithm Calculating a Weighted Average on a Real Sphere

In the previous chapter, a weighted average on a real sphere is defined and has been proven with nice properties. If the uniqueness requirement is satisfied, there is a neighborhood of $\mathbf{p}_1, \dots, \mathbf{p}_n$, $\omega_1, \dots, \omega_n$ in which the weighted average $\mathbf{q}$ is a C^∞-function of $\mathbf{p}_1, \dots, \mathbf{p}_n$, $\omega_1, \dots, \omega_n$. Every point in the convex hull C of $\{\mathbf{x}_1, \dots, \mathbf{x}_k\}$ can be written as a strongly proper weighted average of at most $d+1$ many of $\mathbf{x}_1, \dots, \mathbf{x}_k$. Hence an algorithm of calculating the weighted average on the real sphere is developed. The key idea of this algorithm is defining a logarithmic mapping which maps all points $\mathbf{p}_i$ to the tangent hyperplane at $\mathbf{q}$. Then the Euclidean weighted average $\mathbf{u}$ in the hyperplane is calculated. Finally the result is mapped back to the sphere through the exponential mapping. The mappings mentioned above should preserve the distance and the angle between $\mathbf{p}_i$ and $\mathbf{q}$. The iterative algorithm is described below [5]:

Algorithm calculating the weighted average on the real sphere

- Input: $\mathbf{p}_1, \dots, \mathbf{p}_n$ on S^d and non-negative weights $\omega_1, \dots, \omega_n$ with sum 1;
- Output: the spherical weighted average of the input points;
- Initialization: set $\mathbf{q} := \sum_{i=1}^n \omega_i \mathbf{p}_i / \| \sum_{i=1}^n \omega_i \mathbf{p}_i \|$;
- Main Loop:
  for $i = 1, \dots, n$,
  set $\mathbf{p}_i^* := l_{\mathbf{q}}(\mathbf{p}_i)$,
  set $\mathbf{u} := \sum_{i=1}^n \omega_i (\mathbf{p}_i^* - \mathbf{q})$,

set $\mathbf{q} := \exp_{\mathbf{q}}(\mathbf{q} + \mathbf{u})$,
if $\|\mathbf{u}\|$ is sufficiently small, output $\mathbf{q}$ and halt, otherwise continue looping.

Here $l_{\mathbf{q}}(\mathbf{p}_i)$ is the mapping which maps point $\mathbf{p}_i$ to the tangent hyperplane at $\mathbf{q}$ and $\exp_{\mathbf{q}}(\mathbf{q} + \mathbf{u})$ maps the result back to the sphere. Fig. 5.1 is one example applying this algorithm. Fig. 5.1(a) shows 15 input points on the sphere. The red star on Fig. 5.1(b) shows the weighted average of the input points.

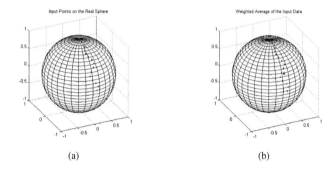

(a)   (b)

Fig. 5.1 Calculating a weighted average on a real sphere: (a) Input points; (b) Weighted average on the sphere.

5.2 Real Spherical Spline Interpolation Algorithm

Using the spherical spline defined in previous chapter, we can solve the spherical spline interpolation problem. Suppose that we are given points $\mathbf{c}_1, \ldots, \mathbf{c}_n$ on the d-sphere, and are given time values $t_1 < t_2 < \ldots < t_n$, we wish to find a smooth curve lying on the sphere, parameterized by t, such that $\mathbf{s}(t_i) = \mathbf{c}_i$ for all i. The basic problem is to choose additional knot positions and control points $\mathbf{p}_i$ defining a spherical spline curve $\mathbf{s}(t) = \widetilde{\sum}_{i=1}^{n} f_i(t)\mathbf{p}_i$ that satisfies these conditions. In this dissertation, we implement the algorithm for the cubic B-spline interpolation, which is an iterative method of solving for the control points $\mathbf{p}_i$. According to the definition, there are $n+2$ control points for n input points. We let $\mathbf{p}_1 = \mathbf{p}_2$ and $\mathbf{p}_{n+1} = \mathbf{p}_{n+2}$, α_i, β_i, γ_i denote the non-zero elements in the basis matrix:

$$\begin{pmatrix} 1 & 0 & 0 & \dots & & 0 & 0 \\ \alpha_2 & \beta_2 & \gamma_2 & 0 & \dots & & 0 \\ 0 & \alpha_3 & \beta_3 & \gamma_3 & 0 & & \vdots \\ \vdots & & \ddots & \ddots & \ddots & & 0 \\ 0 & \dots & 0 & \alpha_{n-1} & \beta_{n-1} & \gamma_{n-1} \\ 0 & 0 & \dots & 0 & 0 & 1 \end{pmatrix} \qquad (5.1)$$

Algorithm for cubic B-spline interpolation:

- Input: points $c_1, \dots, c_n$ and real coefficients $\alpha_i, \beta_i, \gamma_i, (0 \le i \le n)$;
- Output: control points p_i;
- Initialization: set $p_i := c_i$, for $i = 1, \dots n$;
- Main Loop:
  for $i = 1, \dots, n$,
  set $p_{i+1}^* := -(\alpha_i \cdot l_{c_i}(p_i) + \gamma_i \cdot l_{c_i}(p_{i+2}))/\beta_i$,
  set $\delta_i = \|p_i^* - l_{c_i}(p_i)\|$,
  for $i = 1, \dots, n$,
  set $p_{i+1} := \exp_{c_i}(p_i^*)$,
  set $p_1 = p_2$ and $p_{n+2} = p_{n+1}$,
  if the values of δ_i are all sufficiently small, halt; otherwise, continue looping.

When the control points are derived, the spherical B-spline is to be evaluated as spherical weighted averages of the control points. The run time of the spherical weighted average algorithm is one order less than the run time of the interpolation algorithm, so the time to calculate a large number of points along the curve dominates the time needed to calculate the control points. The test results for the cubic B-spline interpolation algorithm are shown in Fig. 5.2.

5.3 Dual Spherical Spline Interpolation

5.3.1 Algorithm calculating a weighted average on the DUS

In this section, a new algorithm to calculate the weighted average on the DUS is proposed. The key idea of this algorithm is to use the logarithmic mapping which maps all points $\hat{p}_i$ on the DUS to the tangent hyperplane at $\hat{q}$, then calculates their weighted average in the hyperplane and maps this result back to the DUS by the exponential mapping. The exponential mapping is defined according to Eq. (4.20) and Eq. (4.21). The logarithmic mapping is defined according to Eq. (4.22). All the calculation rules are based on the calculation rules defined in the dual vector space $\mathbb{D}^3$.

Since the exponential mapping and logarithmic mapping are defined only on $\hat{q} = (0, 0, 1)$, for general point $\hat{q}$ on the DUS, we need to move the coordinate frame in order to fit the mapping. The matrix for moving a point $\hat{x}_1$ to a point $\hat{x}_2$ on the

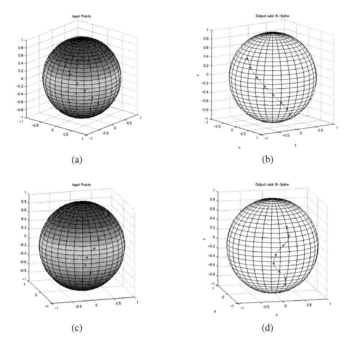

Fig. 5.2 Test result of the cubic B-spline interpolation on the sphere: (a) Given points I; (b) Interpolation result I; (c) Given points II; (d) Interpolation result II.

DUS is given as [55]:

$$\hat{\mathbf{x}}_2 = [\hat{R}]\hat{\mathbf{x}}_1, \tag{5.2}$$

where

$$\begin{aligned}[\hat{R}] &= \exp(\hat{\omega}[ad\hat{\mathbf{g}}]) \\ &= [I] + \sin(\hat{\omega})[ad\hat{\mathbf{g}}] + (\cos(\hat{\omega}) - 1)([ad\hat{\mathbf{g}}])^2, \end{aligned} \tag{5.3}$$

$\hat{\omega}$ equals the dual angle between the points $\hat{\mathbf{x}}_1$ and $\hat{\mathbf{x}}_2$:

$$\hat{\mathbf{x}}_1 \cdot \hat{\mathbf{x}}_2 = \cos\hat{\omega} = x + \varepsilon x^\circ, \tag{5.4}$$

and the screw axis $\hat{\mathbf{g}}$ is chosen to be perpendicular to both points $\hat{\mathbf{x}}_1$ and $\hat{\mathbf{x}}_2$:

$$\hat{\mathbf{g}} = \frac{\hat{\mathbf{x}}_1 \times \hat{\mathbf{x}}_2}{\|\hat{\mathbf{x}}_1 \times \hat{\mathbf{x}}_2\|}. \tag{5.5}$$

The flowchart of the algorithm calculating a weighted average on the DUS is shown in Fig. 5.3, in which $\delta_{INITIAL}$ is the initial error and δ_{TOL} is the error tolerance. $l_{\hat{\mathbf{q}}}(\hat{\mathbf{p}}_i)$ is the mapping which maps point $\hat{\mathbf{p}}_i$ to the tangent hyperplane at $\hat{\mathbf{q}}$ and

$\exp_{\hat{q}}(\hat{q}+\hat{u})$ maps the result back to the DUS. As long as the input points satisfy the uniqueness condition, the algorithm converges.

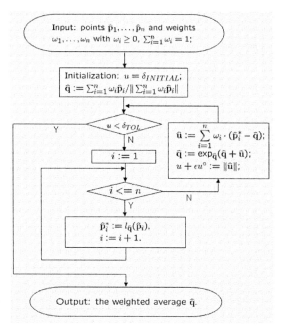

Fig. 5.3 Flowchart of the algorithm calculating a weighted average on the dual unit sphere.

5.3.2 Algorithm of the spline interpolation on the DUS

Adopting the definition in Eq. (4.28), the discrete points on the DUS can be inter-polated as a dual spherical spline. Suppose we are given points $\hat{c}_1$, ..., $\hat{c}_n$ on the DUS and parameters $u_1 < u_2 < ... < u_n$, we wish to find a smooth curve lying on the DUS, parameterized by u, such that $\hat{s}(u_i) = \hat{c}_i$ for all i. The basic problem is to choose additional knot positions and control points $\hat{p}_i$ which define a spherical spline curve $\hat{s}(u) = \widetilde{\sum}_{i=1}^{n} f_i(u)\hat{p}_i$ satisfying these conditions. In this paper, we choose $f_i(u)$ as cubic B-spline basis functions and use an iterative method to solve for the control points $\hat{p}_i$. It can be easily extended to higher order B-splines.

According to the definition, there are $n+2$ control points for n input points. We let $\hat{p}_1 = \hat{p}_2$ and $\hat{p}_{n+1} = \hat{p}_{n+2}$, α_i, β_i, γ_i denote the non-zero elements in the basis matrix Eq. (5.1).

A flowchart of the dual spherical cubic B-spline interpolation algorithm is shown in Fig. 5.4. After the control points are derived, the dual spherical B-spline is to be evaluated as weighted averages of the control points. The run time of the weighted average algorithm is one order less than the run time of the interpolation algorithm, so the time to calculate a large number of points along the curve dominates the time needed to calculate the control points.

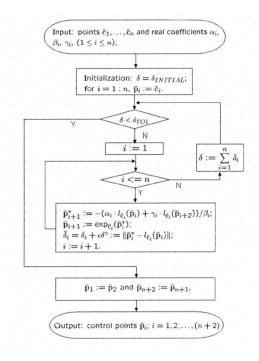

Fig. 5.4 Flowchart of the dual spherical cubic B-spline interpolation algorithm.

5.3.3 Simulation result

We test the algorithm with different inputs. The input line sequence is given in the form of dual vectors $\hat{\mathbf{l}}_i = \mathbf{l}_i + \varepsilon \mathbf{l}_i^\circ$, where $i = 1, \ldots, n$. A point on the DUS corresponds to an infinite line in Euclidean space. In order to display the input line sequence, the dual vector representation of lines is transformed to the algebraic representation of lines:

$$\mathbf{l}_i(v) = \mathbf{l}_i \times \mathbf{l}_i^\circ + v \cdot \mathbf{l}_i, \text{ for } i = 1, \ldots, n \qquad (5.6)$$

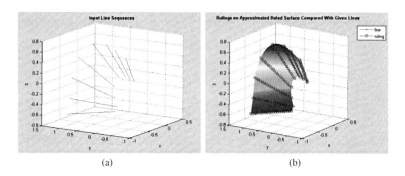

Fig. 5.5 Dual spherical cubic B-spline interpolation I: (a) Input line sequence; (b) Interpolated dual spherical cubic B-spline (ruled surface).

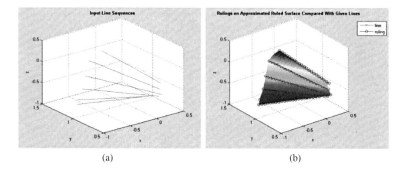

Fig. 5.6 Dual spherical cubic B-spline interpolation II: (a) Input line sequence; (b) Interpolated dual spherical cubic B-spline (ruled surface).

Fig. 5.5(a) and Fig. 5.6(a) show two different input line sequences, where $v \in [0, 1]$. To apply the interpolation algorithm for dual spherical cubic B-splines, the parameters and knot sequence need to be determined. In our program, we choose the chord length to define the parameters: let $\hat{d}_i$ be the chord length between two given points: $\hat{d}_i = \hat{\mathbf{l}}_i \cdot \hat{\mathbf{l}}_{i-1}, i = 1, \ldots, n$, the total chord length is calculated as $\hat{d} = \sum_{i=1}^{n} \hat{d}_i$. Since $\hat{d}_i$ is a dual number, we take the principle part of $\hat{d}_i$ as d_i, the parameters are calculated as:

$$u_0 = 0;$$
$$u_i = u_{i-1} + \frac{d_i}{d}, \text{ for } k = 1, \ldots, n - 1; \qquad (5.7)$$
$$u_n = 1.$$

This dual spherical spline allows the use of arbitrary knot positions. For simplicity, the knots sequence is chosen corresponding to the parameters.

The algorithm converges quickly and the interpolation error is very small. The final result is given as a dual spherical cubic B-spline: $\hat{s}(u) = \sum_{i=1}^{n} f_i(u)\hat{p}_i$, which satisfies $\hat{s}(u_i) = \hat{l}_i$. Fig. 5.5(b) and Fig. 5.6(b) display the dual cubic B-spline as a ruled surface given by Eq. (4.6), where $v \in [0,1]$. The red lines in the figure represent the input line sequence and the blue lines represent the points on the interpolated spline given by $\hat{s}(u_i)$. The red lines and blue lines are overlapping each other, which verifies the algorithm.

5.4 Application in Ruled Surface Approximation

At the beginning of Chapter 4, we have set up the mapping which transforms the ruled surface representation in Euclidean space into a one-parameter curve on the dual unit sphere. The rulings in Euclidean space are corresponding to a sequence of points on the dual unit sphere. The key step for the kinematic approximation of ruled surface is to find a smooth spline on the dual unit sphere which interpolates those points. Finally, the spherical dual spline can be transformed back to a ruled surface in Euclidean space based on Eq. (4.6). Combining all the parts together, we get an algorithm for kinematic ruled surface approximation, it contains the following steps.

Algorithm for ruled surface approximation

- Input: line sequence coordinates in Euclidean space: $\mathbf{P}_i, \mathbf{Q}_i$;
- Output: ruled Surface given by $\mathbf{x}(u,v) = \mathbf{a}(u) + v\mathbf{r}(u)$;
- Main Steps:

  1. transform the coordinates of the line sequences to the coordinates of the points on the DUS,
  2. map points on the DUS to the tangent planes,
  3. interpolate the points with a dual B-spline,
  4. transform the coordinates of the dual B-spline back to the DUS,
  5. evaluate the dual spherical B-spline with the dual spherical weighted average algorithm,
  6. transform the dual number representation of ruled surface back to the Euclidean space.

We test the algorithm with different inputs. The ruling sequences are the the the first step output of the approximation algorithm introduced in Chapter 3. Fig. 5.7 shows two examples used to test the algorithm.

Applying the algorithm for dual spherical cubic B-spline interpolation algorithm, a dual spherical cubic B-spline is derived, which corresponds to a ruled surface in Euclidean space. This curve can be mapped back to a ruled surface in Euclidean space based on Eq. (4.6). Two directrix curves of a ruled surface can be written as Eq. (4.7), which involve two additional parameters δ_{p_1} and δ_{p_2}. Extra information

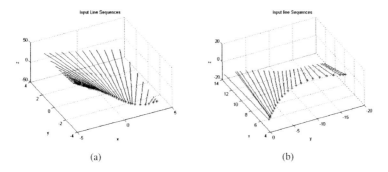

Fig. 5.7 Input line sequences (a) Input line sequences I; (b) Input line sequences II.

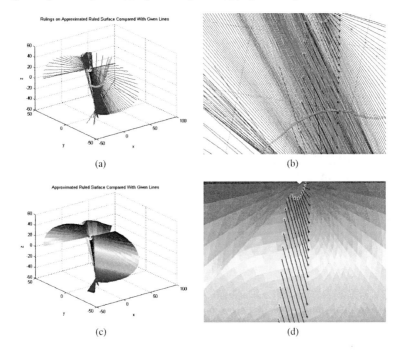

Fig. 5.8 Test result I of the kinematic ruled surface approximation: (a) Comparison of rulings on the ruled surface with input lines; (b) Comparison of rulings on the ruled surface with input lines; (c) Comparison of the ruled surface with input lines; (d) Details of comparison of the ruled surface with input lines.

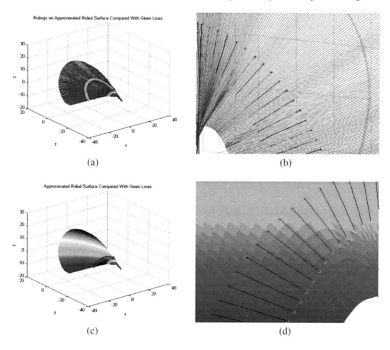

(a)   (b)

(c)   (d)

Fig. 5.9 Test result II of the kinematic ruled surface approximation: (a) Comparison of rulings on the ruled surface with input lines; (b) Comparison of rulings on the ruled surface with input lines; (c) Comparison of the ruled surface with input lines; (d) Details of comparison of the ruled surface with input lines.

is needed to decide the boundary of the ruled surface. In this chapter, we simply choose δ_{p_1} and δ_{p_2} as constant to display the ruled surface. Fig. 5.8 and Fig. 5.9 compare the interpolation results with the input line sequences. From the figures, we can see that the approximated ruled surface fits well to the given line sequence, though the boundary of the ruled surface does not describe the boundary character of the given surface.

5.5 Comments

In this chapter, we first implement the algorithm calculating the weighted average on the real sphere and the spherical spline interpolation algorithm. Then, a novel algorithm for weighted average on DUS calculation and an algorithm of dual spherical spline interpolation are proposed. These algorithm can be applied for ruled surface approximation. Since a point on the dual unit sphere corresponds to an infinite line in Euclidean space, we need extra information to decide the boundary of the ruled surface. In this chapter, the ruled surface is determined based on the mapping defined by Eq. (4.5). It is to find the point D on the ruling which OD is perpendicular to the ruling. Then two directrix curves of a ruled surface can be written as:

$$\mathbf{r}_{p_1}(u) = \mathbf{r}_D(u) + \delta_{p_1}(u)\mathbf{l}(u); \tag{5.8a}$$

$$\mathbf{r}_{p_2}(u) = \mathbf{r}_D(u) + \delta_{p_2}(u)\mathbf{l}(u). \tag{5.8b}$$

However, from the simulation results, we see that it has little control with the point D and requires two more parameters to store for each ruling. In the next chapter, we are going to propose an improved algorithm for kinematic ruled surface approximation, which has more control about the directrix curves. The algorithm proposed in the next chapter also provides more information about the approximation error.

Chapter 6
Kinematic Ruled Surface Approximation Algorithm

In this chapter, we propose an improved kinematic ruled surface approximation algorithm which has more boundary control. We know that a point can be interpreted as an intersection of two lines, so a point on the boundary of a ruled surface can be defined by intersecting a ruling with a reference line. In order to be precise, the reference line is defined passing through a point on a directrix curve and the orientation of the reference line is consistent with the surface normal on that point. This definition of the reference line is inspired by the manufacturing process, where the orientation of the generator (ruling), the normal of the surface and the bi-normal that perpendicular to both ruling and the normal constitute a coordinate system for the moving milling machine. Obviously, the movement of the reference line also produces a ruled surface. Therefore, we can apply the dual spherical spline interpolation algorithm using the reference lines as input. One directrix curve is derived by intersecting those two ruled surfaces. Similarly, we can derive the other directrix curve by repeating the previous procedure.

6.1 Framework of the Kinematic Ruled Surface Approximation Algorithm

In this section, we first propose a framework of the kinematic approximation algorithm. Based on the algorithm introduced in the previous chapter, we add two more steps to get the complete framework [66]:

1. Extract the rulings and determine the reference lines
2. Write the coordinates of the lines as dual vectors
3. Apply the dual spherical B-spline interpolation algorithm
4. Evaluate the dual spherical B-spline
5. Transform the ruled surface back to the Euclidean space (optional)

In the following sections, we are going to explain the new added steps and show the effect of the boundary control.

6.1.1 Rulings and reference lines extraction

For a given surface, the first step is to extract a sequence of rulings from the surface. Here, we adopt the similar procedure as the first step of the algorithm introduced in Chapter 3. We should notice that it has a condition that there exists a vector e_3 and an angle $0 < \gamma_0 < \pi/2$, such that all surface normals form an angle $0 < \gamma < \gamma_0$. In certain cases, we need to change the coordinate system in order to satisfy this condition.

In Chapter 3, the middle point M_0 of domain D' is chosen as a starting point and the search procedure needs to be carried out in two directions. In our algorithm, we start the search from the leading edge of the given blade to the tailing edge. This change is to fit the manufacturing technique, in which the milling tool usually moves in one direction.

Besides the ruling sequences, the reference lines are also needed to be determined in this step. As introduced above, the reference line should pass through the point on the directrix curve and parallel to the surface normal on that point. Since the two directrix curves should pass the ending points of the rulings, once the rulings are derived, the intersection point between the ruling and the directrix curve is derived instantly. The direction of the reference line is determined by the cross product of the ruling directional vector and the tangent vector of the directrix curve[1]. In the end, we get three sequences of lines, which contain one group of rulings and two groups of reference lines. Using these line sequences as input, we can apply the dual spherical interpolation algorithm introduced in the previous chapter to get three ruled surfaces.

6.1.2 Line intersection

As we know, the movement of a line in Euclidean space is equivalent to the movement of a point on the dual unit sphere. The properties of the displacement matrices, which are used to generate a curve on the DUS, preserve the incidence relationship between lines. If $\hat{\mathbf{x}}_1$ and $\hat{\mathbf{y}}_1$ denote two lines, then the scalar product between them will be preserved after the displacement. Let $\hat{\mathbf{x}}_2 = [\hat{R}]\hat{\mathbf{x}}_1$ and $\hat{\mathbf{y}}_2 = [\hat{R}]\hat{\mathbf{y}}_1$, since the displacement matrices are orthogonal matrices, we get:

$$(\hat{\mathbf{x}}_2)^T \hat{\mathbf{y}}_2 = (\hat{\mathbf{x}}_1)^T [\hat{R}]^T [\hat{R}]\hat{\mathbf{y}}_1 = (\hat{\mathbf{x}}_1)^T \hat{\mathbf{y}}_1. \tag{6.1}$$

This means that two lines that intersect each other will still be intersecting after the displacement.

Using this property, we can find a point trajectory from a displacement of two intersecting lines. The directrix curve is thought as a point trajectory, which is an intersection of a ruling and a reference line. Then applying the interpolation algorithm to

---

[1] Here we use the approximation of the tangent vector since the directrix has not been determined.

the reference lines and rulings respectively, using the same parametrization method, we derive the displacement matrix for both rulings and reference lines. Since the scalar product is preserved, the reference lines will always intersect the rulings at the end of the algorithm. The intersection points can be used as the boundary curve for the ruled surface.

Now, we introduce an algorithm which calculates the intersection point of two lines [55]. The input for this algorithm are two sets of Plücker coordinates of two lines $(\mathbf{u}_1, \mathbf{u}_1^\circ)$ and $(\mathbf{u}_2, \mathbf{u}_2^\circ)$. Rewrite the Plücker coordinates in the parametric form of lines:

$$\mathbf{l}_1(t) = \mathbf{u}_1 \times \mathbf{u}_1^\circ + \mathbf{u}_1 t = \mathbf{c}_1 + \mathbf{u}_1 t, \tag{6.2a}$$

$$\mathbf{l}_2(t) = \mathbf{u}_2 \times \mathbf{u}_2^\circ + \mathbf{u}_2 t = \mathbf{c}_2 + \mathbf{u}_2 t. \tag{6.2b}$$

If these two lines intersect each other, the intersection point should satisfy the equation $\mathbf{l}_1(t_1) = \mathbf{l}_2(t_2)$. Subtract Eq. (6.2a) and Eq. (6.2b):

$$\mathbf{c}_1 - \mathbf{c}_2 = [\mathbf{u}_2 \; -\mathbf{u}_1] \begin{pmatrix} t_2 \\ t_1 \end{pmatrix}. \tag{6.3}$$

Eq. (6.3) has the form $\mathbf{b} = A\mathbf{x}$, where A is a 3×2 matrix. This system can be solved by multiplying both side of Eq. (6.3) by A^T:

$$A^T \mathbf{b} = A^T A \mathbf{x}. \tag{6.4}$$

Then we get the following system:

$$\begin{pmatrix} \mathbf{u}_2 \mathbf{c}_1 \\ \mathbf{u}_1 \mathbf{c}_2 \end{pmatrix} = \begin{pmatrix} 1 & -\mathbf{u}_1^T \mathbf{u}_2 \\ -\mathbf{u}_1^T \mathbf{u}_2 & 1 \end{pmatrix} \begin{pmatrix} t_2 \\ t_1 \end{pmatrix}. \tag{6.5}$$

This system is easily solved for t_1 and t_2. If the two lines intersect each other, there is only one solution and it is the minimum norm solution of Eq. (6.3).

For the case that two lines are skew, the results of the previous algorithm give two points located on each line respectively that the distance between them is the minimum distance between two lines. It can be proven in a simple way.

Suppose a point on the line $\mathbf{l}_1$ is denoted as $\mathbf{l}_1(t_1)$ and a point on the line $\mathbf{l}_2$ is denoted as $\mathbf{l}_2(t_2)$, the distance between these two points are the length of the vector $\mathbf{v} = \mathbf{l}_1(t_1) - \mathbf{l}_2(t_2)$. To find the minimum distance between the lines, we should minimize the square length of the vector $\mathbf{v}$:

$$\begin{aligned} f &= \mathbf{v}^T \mathbf{v} \\ &= (\mathbf{l}_1(t_1) - \mathbf{l}_2(t_2))^T (\mathbf{l}_1(t_1) - \mathbf{l}_2(t_2)) \\ &= \mathbf{l}_1(t_1) \cdot \mathbf{l}_1(t_1) - 2\mathbf{l}_1(t_1) \cdot \mathbf{l}_2(t_2) + \mathbf{l}_2(t_2) \cdot \mathbf{l}_2(t_2) \\ &= (\mathbf{c}_1 \cdot \mathbf{c}_1) + t_1^2 + (\mathbf{c}_2 \cdot \mathbf{c}_2) + t_2^2 \\ &\quad -2[(\mathbf{c}_1 \cdot \mathbf{c}_2) + (\mathbf{c}_1 \cdot \mathbf{u}_2)t_2 + (\mathbf{c}_2 \cdot \mathbf{u}_1)t_1 + (\mathbf{u}_1 \cdot \mathbf{u}_2)t_1 t_2]. \end{aligned} \tag{6.6}$$

The values of t_1 and t_2 which minimize the objective functional f should satisfy $\frac{\partial f}{\partial t_1} = \frac{\partial f}{\partial t_2} = 0$:

$$\frac{\partial f}{\partial t_1} = 2t_1 - 2(\mathbf{c}_2 \cdot \mathbf{u}_1) - 2(\mathbf{u}_1 \cdot \mathbf{u}_2)t_2, \tag{6.7a}$$

$$\frac{\partial f}{\partial t_2} = 2t_2 - 2(\mathbf{c}_1 \cdot \mathbf{u}_2) - 2(\mathbf{u}_1 \cdot \mathbf{u}_2)t_1. \tag{6.7b}$$

After simplifying these expressions, we get:

$$\begin{pmatrix} \mathbf{u}_2\mathbf{c}_1 \\ \mathbf{u}_1\mathbf{c}_2 \end{pmatrix} = \begin{pmatrix} 1 & -\mathbf{u}_1^T\mathbf{u}_2 \\ -\mathbf{u}_1^T\mathbf{u}_2 & 1 \end{pmatrix} \begin{pmatrix} t_2 \\ t_1 \end{pmatrix}. \tag{6.8}$$

This is exactly the same solution given by Eq. (6.3). In short, the minimum norm solution gives the points on lines whose distance is the minimum distance between two lines.

This algorithm will fail to find a solution if two lines are parallel or coincident. In those two cases, the matrix $A^T A$ will be singular and all points will be at the minimum distance from the corresponding points on the other line. However, in the beginning, we have ruled out this situation by setting the reference line perpendicular to the ruling, we can apply this algorithm to evaluate the boundary of the approximated ruled surface.

6.2 Application in Turbocharger Blade Design

In this section, the algorithm is applied to approximate a given turbocharger blade by a ruled surface. A turbocharger blade is a solid model containing the pressure and suction surfaces. Here, we approximate its two sides respectively.

Now we test this algorithm with different input blades and study the approximation error. Fig. 6.1 shows the simulation result of a pressure surface of a given blade. Fig. 6.1(a) contains the rulings and reference lines extracted from the given blade surface. The black lines are the rulings and the green and blue lines represent the reference lines for two directrix curves. The intersection points of the reference lines and the rulings are also drawn with different colors. Using those line sequences as input, we can apply the kinematic ruled surface approximation algorithm. In the end, the ruled surface is evaluated by the dual weighted average algorithm. Two boundaries of the ruled surface are determined by the line intersection algorithm. Fig. 6.1(c) compares the approximated ruled surface with the given blade, the red surface is the approximated ruled surface and the colored surface is the given blade, we can see that these two surfaces are very close to each other. Actually, the average approximation error is only 0.0062mm and the maximum approximation error is 0.1129mm. Therefore, the kinematically approximated ruled surface can be used instead of the original design.

Fig. 6.2 shows the simulation result for the suction surface of the given blade. From these figures, we can see that the approximation error for the other side of the given blade is also very small. For this case, the average approximation error is 0.0052mm and the maximal approximation error is 0.1207mm.

The approximation error is related to the original shape of the given blade. For the above example, the given blade is close to a ruled surface, therefore, the approximation error is small. However, for some examples, the approximation error will be bigger. Fig. 6.3 shows one of those examples. The tail part of the given blade is convex, where the approximation error is much bigger than other parts.

From the simulation examples, we also find that the distribution of the extracted rulings is also important for the performance of the algorithms. Fig. 6.4 and Fig. 6.5 show the influence of the ruling location on the approximation error. In these two figures, the test blade is the same, only the distributions of approximation rulings are different. Obviously, the even distribution of the approximation rulings leads to better approximation result.

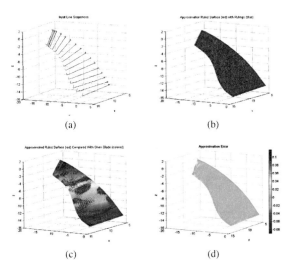

(a)   (b)

(c)   (d)

Fig. 6.1 Approximation result of the given blade I (pressure surface): (a) Rulings and reference lines extracted from the given blade; (b) Kinematic approximated ruled surface; (c) Comparison of the ruled surface with the given blade; (d) Approximation error.

6.3 Comments

In this chapter, we propose a new kinematic ruled surface approximation algorithm which has more control on the boundary. This algorithm is set up based on the manufacturing procedure and employs the basic property in line geometry. We have known that the intersection property preserves during the displacement. Using this

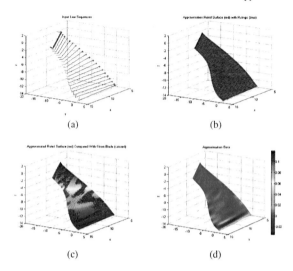

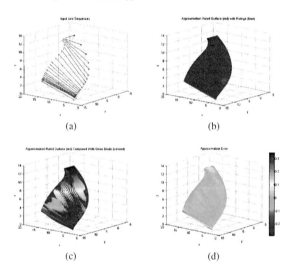

Fig. 6.2 Approximation result of the given blade I (suction surface): (a) Rulings and reference lines extracted from the given blade; (b) Kinematic approximated ruled surface; (c) Comparison of the ruled surface with the given blade; (d) Approximation error.

Fig. 6.3 Approximation result of the given blade II (suction surface): (a) Rulings and reference lines extracted from the given blade; (b) Kinematic approximated ruled surface; (c) Comparison of the ruled surface with the given blade; (d) Approximation error.

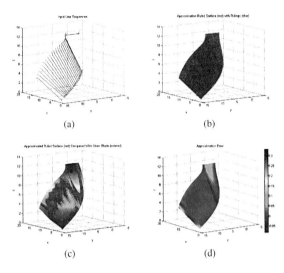

Fig. 6.4 Approximation result I of the given blade II (pressure surface): (a) Rulings and reference lines extracted from the given blade; (b) Kinematic approximated ruled surface; (c) Comparison of the ruled surface with the given blade; (d) Approximation error.

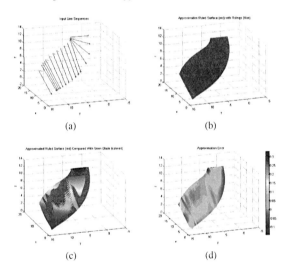

Fig. 6.5 Approximation result II of the given blade II (pressure surface): (a) Rulings and reference lines extracted from the given blade; (b) Kinematic approximated ruled surface; (c) Comparison of the ruled surface with the given blade; (d) Approximation error.

property, a line intersection algorithm is used to determine the boundary of the ruled surface.

We test this algorithm with different blade surfaces. In most cases, the algorithm gives good approximation result. The approximation error is very small and the approximated ruled surface can be used instead of the original blade design. We also find that the distribution of the rulings and reference lines are very important for the performance of the algorithm. If the rulings are too sparse or distribute unevenly, it will increase the approximation error.

This algorithm has many applications in turbocharger blade design, which approximates the original free-form surface with a rule surface. Since ruled surface is easy and cheap to manufacture, the manufacturing cost is decreased. However, the approximation error relies strongly to the shape of the original design. If the original design is far away from ruled surface, it is difficult to find a good approximation to replace the original design. In next chapter, an optimization prototype for ruled surface is set up. A ruled surface can be defined by several control points of the dual spherical spline. Through adjusting these parameters, engineers can design an optimal ruled surface under the constrains of design requirements.

Chapter 7
A Prototype for Optimization

In this chapter, we try to set up an initial prototype for the ruled surface optimization. In this model, a few parameters are extracted to define a ruled surface. Those parameters are optimized to satisfy certain functional requirements. Normally, the number of parameters should be as less as possible to reduce the complexity of the optimization process.

In Chapter 4, we have already known that a ruled surface can be represented as a curve on the DUS, in which a novel dual spherical spline is defined. Using the dual spherical interpolation algorithm, a given surface is approximated with a ruled surface, which is defined by a sequence of control points in the form of dual vectors. Each dual vector contains six elements, however satisfying two constrains (Plücker condition and unit length condition). Therefore it has four degrees of freedom. Inspired by the exponential and logarithmic mappings defined in Chapter 4, we can use the control points on a tangent plane as the optimization parameters. Compared with the conventional way using control points of two directrix curves as optimization parameters, it can save one third number of parameters.

Since each point on the DUS corresponds to an infinite line in Euclidean space, extra information is needed to define the rulings of a ruled surface as line segments. In specific blade geometry optimization problems, the hub and shroud surfaces are fixed. So the two directrix curves of a ruled surface can be interpreted as two intersection curves between the ruled surface and hub/shroud surface. Therefore, combining the dual spherical interpolation algorithm with an algorithm that calculates the intersection curve between a rule surface and a free-form surface, we can determine a ruled surface by several control points of a dual spherical spline. It provides an initial prototype for the blade geometry optimization with ruled surface.

In order to control the number of parameters, the dual spherical spline interpolation algorithm is extended to the dual spherical spline approximation algorithm. Using this algorithm, the number of the parameters is determined at the initialization phase.

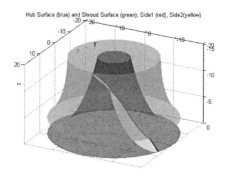

Fig. 7.1 A given blade model.

7.1 A Prototype for Optimization

In this section, we first draw a rough picture of the optimization prototype. This optimization prototype is proposed especially for the blade geometry optimization, which adapts to the commonly used blade geometry design software "Bladegen".

In Chapter 3, we show some examples of products with blade components. Mathematically, a blade can be abstracted as two surfaces (pressure surface and suction surface), which represent two sides of a blade respectively. Fig. 7.1 shows a mathematical model of a blade. The hub and shroud surfaces are generated by rotating hub and shroud curves along the central axis respectively. In order to reduce the manufacturing cost, the pressure and suction surfaces are usually designed as two ruled surfaces.

Based on this blade model, two directrix curves of the ruled surface, which are also called hub and shroud curves, can be interpreted as the intersection curves between the rule surface and the hub and shroud surfaces. Hence, we can set up an optimization prototype for the blade geometry design. Fig. 7.2 shows the diagram of this prototype.

In the above diagram, the input files are pulled out from "Bladegen" software. The "ibl" file contains point-wise information of a given blade and the "bgi" file contains the layer information as well as the thickness and angle information for a given blade. First, we use the dual spherical spline interpolation algorithm to get an approximated ruled surface from a given "ibl" file. Simultaneously, we abstract the information of hub and shroud surfaces from the corresponding "bgi" file. Then the intersection algorithm is applied to construct the blade surface. The control points of the dual spherical spline are mapped to a tangent plane to get the parameters for the optimizer. Each time the optimizer changes the parameters, it corresponds to a new dual spherical spline and can construct a new blade surface. This new blade surface

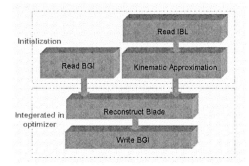

Fig. 7.2 A diagram of the blade geometry optimization prototype.

is evaluated by certain functional requirements and guides the change of parameters in the next step.

7.2 Intersect a Ruled Surface with a Free-form Surface

For the surface-surface intersection problems, there are plenty of algorithms developed to solve this type of problems [34] [28] [26] [22] [60]. For our special application case, we choose an algorithm intersecting a ruled surface with a free-form surface, which is proposed in [60], to determine the hub and shroud curves. This algorithm employs the special properties of ruled surface to reduce the calculation complexity. It is easy to implement and fast enough to be integrated in the optimization loops. In this section, we introduce the essence of this intersection algorithm and test it with simple examples.

7.2.1 Algorithm description

Generally, there are two types of representations for a free-form surface: parametric representation and implicit representation. Considering the practical application, the blade geometry is determined by certain control points of Bézier curves. Therefore, we only focus on the parametric representation. The free-form surface is denoted as a parametric surface described by a vector-valued function of two variables:

$$\mathbf{S}(u,v) = (x(u,v),y(u,v),z(u,v)); \quad (u,v) \in D \subset \mathbb{R}^2, \tag{7.1}$$

where D is the surface domain and $x(u,v),y(u,v),z(u,v)$ are functions of variables u and v. Usually, the surface is differentiable and represented as a tensor product

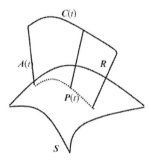

Fig. 7.3 A model of intersecting a ruled surface with a free-form surface.

surface. As stated in Chapter 2, a ruled surface in $\mathbb{R}^3$ has a parametric representation:

$$\mathbf{R}(t,s) = \mathbf{C}(t) + s\mathbf{A}(t);\ t \in I, s \in \mathbb{R}, \tag{7.2}$$

where $\mathbf{C}(t)$ denotes a directrix curve and $\mathbf{A}(t)$ denotes a vector field.

If a ruled surface $\mathbf{R}$ and a free-form surface $\mathbf{S}$ intersect transversely, the intersection curve is the projection curve of directrix curve $\mathbf{C}(t)$ onto $\mathbf{S}$ along the vector field $\mathbf{A}(t)$ [60]. Using this method, the intersection curve is calculated by solving a system of Ordinary Differential Equations (ODEs). The intersection curve is denoted as $\mathbf{P}(t) = \mathbf{S}(u,v)$. Obviously $[\mathbf{C}(t) - \mathbf{P}(t)]$ represents the same vector field as $\mathbf{A}(t)$, which can be seen from Fig. 7.3 [1]. Hence we get the following relationship:

$$[\mathbf{C}(t) - \mathbf{P}(t)] \times \mathbf{A}(t) = \mathbf{0}. \tag{7.3}$$

Taking the derivative of Eq. (7.3) with respect to t and using the relation $\mathbf{P}'(t) = \mathbf{S}_u \frac{du}{dt} + \mathbf{S}_v \frac{dv}{dt}$, we get:

$$\mathbf{S}_u \times \mathbf{A}(t) \frac{du}{dt} + \mathbf{S}_v \times \mathbf{A}(t) \frac{dv}{dt} = \mathbf{C}'(t) \times \mathbf{A}(t) + (\mathbf{C}(t) - \mathbf{P}(t)) \times \mathbf{A}'(t). \tag{7.4}$$

Calculating the dot product of the vectors $\mathbf{S}_u$ and $\mathbf{S}_v$ with both side of Eq. (7.4) respectively, we get two equations:

$$\begin{cases} \frac{du}{dt} = \frac{[\mathbf{A}(t) \times \mathbf{C}'(t) - (\mathbf{C}(t) - \mathbf{P}(t)) \times \mathbf{A}'(t)] \cdot \mathbf{S}_v}{[\mathbf{S}_u \times \mathbf{S}_v] \cdot \mathbf{A}(t)}; \\ \frac{dv}{dt} = \frac{[\mathbf{C}'(t) \times \mathbf{A}(t) + (\mathbf{C}(t) - \mathbf{P}(t)) \times \mathbf{A}'(t)] \cdot \mathbf{S}_u}{[\mathbf{S}_u \times \mathbf{S}_v] \cdot \mathbf{A}(t)}; \end{cases} \tag{7.5}$$

where $(\mathbf{S}_u \times \mathbf{S}_v) \cdot \mathbf{A} \neq 0$. Now the problem becomes to solve an ODE system for the given initial values:

---

[1] This figure is taken from [60]

$$\begin{cases} u(t_0) = u_0, \\ v(t_0) = v_0. \end{cases} \tag{7.6}$$

Theoretically, the intersection curve may contain several disconnected segments, those projective curve segments can be computed by choosing different initial values and parametric intervals. Fortunately, in our application, the boundary of the blade surface is a continuous curve, the integration interval is $[0,1]$. For the start point, it is a problem to find the intersection point between a ruling and the free-form surface. In other words: $\mathbf{S}(u,v) - \mathbf{r}(t_0) = \mathbf{0}$. This problem can be solved using the Newton method.

7.2.2 Numerical integration and application

After setting up the ODE system and obtaining the initial points, we can solve the ODEs to find the intersection curve. Although the first-order ODEs presented in the previous section can not be solved analytically except for some special cases, they can be solved efficiently by numerical techniques. For simulation purpose, we use the ODEs solver function "ode45" provided in Matlab to solve the problem, which combines 4th- and 5th-order Runge-Kutta methods for controlling the error and the steps. The output results of "ode45" are discrete points on the intersection curve, which should be interpolated/approximated with a B-spline curve for the practical use.

Here, we implement the algorithm and test it with a simple example provided in [60]. The ruled surface is a hyperboloid of resolution:

$$\mathbf{R}(t,v) = (\sqrt{2}\cos(t), \sqrt{2}\sin(t), 0) + v(-\sqrt{2}\sin(t), \sqrt{2}\cos(t), 4), \tag{7.7}$$

where $t \in [0, 2\pi], v \in [-1,1]$. The free-form surface is a Bézier tensor product surface $\mathbf{S}(x,y)$, whose control points are:

$$\begin{pmatrix} (-4,4,-1) & (0,4,3) & (4,4,-1) \\ (-4,0,4) & (0,0,6) & (4,0,3) \\ (-4,-4,-1) & (0,-4,3) & (4,-4,-1) \end{pmatrix}. \tag{7.8}$$

Fig. 7.4 shows the simulation result for this example. The green surface is a ruled surface and the red surface is a free-form surface.

The initial point $(1.4142, 0.8655, 2.4479)$ is calculated by intersecting the ruling at $t = 0$ of the ruled surface with the free-form surface. The integration range is $[0, 2\pi]$. The intersection curve is plotted with blue color in Fig. 7.4(c).

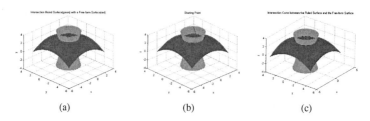

(a)   (b)   (c)

Fig. 7.4 Simulation result of intersecting a ruled surface with a free-form surface: (a) Two given surfaces; (b) Initial point; (c) The intersection curve.

7.3 Initialization of the Optimizer

In this section, we use a real example to explain the initialization process. This blade design is pulled out directly from "Bladegen". Fig. 7.5(a) shows the geometry of the blade. For simplicity, we only show one side of the blade as example. The initialization of the other side follows the similar procedure.

First, the information of the hub and shroud surfaces is abstracted, which are defined by some Bézier curve control points of the top and bottom "Meridional" curves. the top and bottom "Meridional" curves are the projections of the blade top and bottom layers on the $r - z$ plane. The hub and shroud surfaces are generated by rotating these curves along the axis. The reconstructed hub and shroud surfaces are plotted in Fig. 7.5(b). The given blade surface is approximated with a ruled surface using the dual spherical spline interpolation algorithm, which follows the same procedure in Chapter 5. The approximated ruled surface together with the hub and shroud surfaces are plotted in Fig. 7.5(c). Then, choosing the ruling of $t = 0$, we get the initial points for the ODE system by intersecting the ruling with the hub and shroud surfaces. From Fig. 7.5(d), we can see that the initial points coincide with the starting points of two directrix curves. Finally, the two boundaries of the ruled surface are derived by solving the ODE system. Fig. 7.5(e) plots the constructed blade model which is defined by a sequence of dual spherical spline control points. Comparing the constructed blade with the given blade, we can see that the error is small enough to use the constructed blade as an initial shape for the optimizer.

Table. 7.1 lists the control points for the dual spherical spline, which corresponds to the constructed ruled surface. Applying the logarithmic mapping to the control points with respect to its interpolated points, we get a sequence of control points on the corresponding tangent planes. Since the first and last rulings of ruled surface usually remain unchanged. we are only interested in the remaining control points. Table. 7.2 shows the results after mapping, those parameters are sent to the optimizer for the next step. Every time those parameters have been changed, they should be mapped back to the DUS and construct a new ruled surface. The number of the control points can be controlled at the initialization step by controlling the input ruling sequence. However, the reduction of the input ruling number will also cause an increase in the approximation error. Hence, in the next section, we try to develop a dual spherical spline approximation algorithm, which minimizes the distance between the spline the given points globally. The number of the control points can be controlled in advance.

Control Point 1	$(-0.6781, -0.3070, 0.6678) + \varepsilon(-12.5857, -4.4701, -14.8346)$
Control Point 2	$(-0.6781, -0.3070, 0.6678) + \varepsilon(-12.5857, -4.4701, -14.8346)$
Control Point 3	$(-0.6603, -0.2323, 0.7142) + \varepsilon(-12.6778, -4.9245, -13.3229)$
Control Point 4	$(-0.6460, -0.1474, 0.7490) + \varepsilon(-12.5694, -5.2565, -11.8751)$
Control Point 5	$(-0.6214, -0.0047, 0.7835) + \varepsilon(-12.0041, -5.5701, -9.5547)$
Control Point 6	$(-0.6017, 0.2325, 0.7641) + \varepsilon(-9.9372, -5.2619, -6.2244)$
Control Point 7	$(-0.5812, 0.2550, 0.7728) + \varepsilon(-8.4007, -4.9827, -4.6741)$
Control Point 8	$(-0.5955, 0.5619, 0.5742) + \varepsilon(-3.0856, -2.3691, -0.8819)$
Control Point 9	$(-0.6548, 0.6240, 0.4263) + \varepsilon(0.5145, 0.3638, 0.2577)$
Control Point 10	$(-0.7359, 0.6104, 0.2930) + \varepsilon(3.0686, 3.3975, 0.6280)$
Control Point 11	$(-0.8394, 0.4917, 0.2316) + \varepsilon(3.6624, 6.2052, 0.0995)$
Control Point 12	$(-0.9066, 0.4099, 0.1000) + \varepsilon(4.2728, 9.4087, 0.1684)$
Control Point 13	$(-0.9655, 0.2540, 0.0573) + \varepsilon(3.0389, 11.6171, -0.3008)$
Control Point 14	$(-0.9830, 0.1828, -0.0160) + \varepsilon(2.4887, 13.3654, -0.1607)$
Control Point 15	$(-0.9976, -0.0570, 0.0402) + \varepsilon(-0.8352, 13.7916, -1.1859)$
Control Point 16	$(-0.9976, -0.0570, 0.0402) + \varepsilon(-0.8352, 13.7916, -1.1859)$

Table 7.1 Control points of the dual spherical cubic B-spline.

Control Point 3	$(-0.0005, -0.0082, 0) + \varepsilon(-0.0700, -0.0023, 0)$
Control Point 4	$(0.0054, 0.0139, 0) + \varepsilon(0.1653, -0.0142, 0)$
Control Point 5	$(0.0109, 0.0151, 0) + \varepsilon(0.2038, -0.0612, 0)$
Control Point 6	$(0.0013, 0.0506, 0) + \varepsilon(0.4552, -0.0286, 0)$
Control Point 7	$(0.0186, -0.0437, 0) + \varepsilon(-0.2786, -0.1861, 0)$
Control Point 8	$(0.0065, 0.0474, 0) + \varepsilon(0.5969, -0.1099, 0)$
Control Point 9	$(0.0049, 0.0118, 0) + \varepsilon(0.1888, -0.0763, 0)$
Control Point 10	$(0.0029, 0.0223, 0) + \varepsilon(0.3320, -0.0428, 0)$
Control Point 11	$(0.0148, -0.0093, 0) + \varepsilon(-0.2532, -0.2918, 0)$
Control Point 12	$(0.0006, 0.0239, 0) + \varepsilon(0.3725, -0.0243, 0)$
Control Point 13	$(0.0095, -0.0060, 0) + \varepsilon(-0.1908, -0.2396, 0)$
Control Point 14	$(-0.0190, 0.0469, 0) + \varepsilon(0.9211, 0.4218, 0)$

Table 7.2 Control points on corresponding tangent planes.

7.4 Dual Spherical Spline Approximation

In this section, the dual spherical spline interpolation algorithm is extended to a dual spherical spline approximation algorithm. We start with the real spherical spline interpolation algorithm and extend it to a real spherical spline approximation algo-

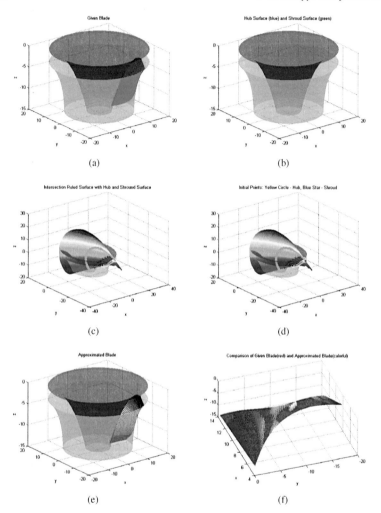

(a)

(b)

(c)

(d)

(e)

(f)

Fig. 7.5 Initialization procedure: (a) Given blade; (b) Hub and shroud surface; (c) Derived ruled surface after using the dual spherical spline interpolation algorithm; (d) Initial points; (e) Constructed blade by intersecting the ruled surface with hub and shroud surface; (f) Comparison between the given and constructed blade.

rithm. Different inputs are used to verify this algorithm and discuss its property. In the end, this algorithm is extended to attain a dual spherical spline approximation algorithm.

7.4.1 Real spherical spline approximation algorithm

Similar as the real spherical spline interpolation algorithm, the real spherical spline approximation is developed based on a mapping between the real sphere and a tangent plane. Here, we choose the tangent plane on the middle point of the given input point sequence. First, this point should be mapped to the point $(0,0,1)$ by rotating the coordinate system. In $\mathbb{R}^3$, coordinate system rotations of the $x-$, $y-$ and $z-$axis in a counterclockwise direction when looking towards the origin give the matrices [3]:

$$R_x(\alpha) = \begin{pmatrix} 1 & 0 & 0 \\ 0 & \cos\alpha & \sin\alpha \\ 0 & -\sin\alpha & \cos\alpha \end{pmatrix}; \tag{7.9}$$

$$R_y(\beta) = \begin{pmatrix} \cos\beta & 0 & -\sin\beta \\ 0 & 1 & 0 \\ \sin\beta & 0 & \cos\beta \end{pmatrix}; \tag{7.10}$$

$$R_z(\gamma) = \begin{pmatrix} \cos\gamma & \sin\gamma & 0 \\ -\sin\gamma & \cos\gamma & 0 \\ 0 & 0 & 1 \end{pmatrix}. \tag{7.11}$$

Any rotation can be given as a composition of rotations about three axes (Euler's rotation theorem), and thus can be represented by a 3×3 matrix operating on a vector:

$$\begin{pmatrix} x'_1 \\ x'_2 \\ x'_3 \end{pmatrix} = \begin{pmatrix} a_{11} & a_{12} & a_{13} \\ a_{21} & a_{22} & a_{23} \\ a_{31} & a_{32} & a_{33} \end{pmatrix} \begin{pmatrix} x_1 \\ x_2 \\ x_3 \end{pmatrix}. \tag{7.12}$$

Using Eq. (7.12), the coordinate system is changed in order to fit the mapping defined based on the point $(0,0,1)$. The points on the tangent plane of point $(0,0,1)$ are mapped to the real sphere by the exponential mapping:

$$\begin{aligned} x'_1 &= x_1 \cdot \frac{\sin(r)}{r}, \\ x'_2 &= x_2 \cdot \frac{\sin(r)}{r}, \\ x'_3 &= \cos(r), \end{aligned} \tag{7.13}$$

where $r = \sqrt{x_1^2 + x_2^2}$. In case $r = 0$, we assign $x'_1 = x_1$ and $x'_2 = x_2$.

Reversely, the points on the real sphere are mapped to the tangent plane by the logarithmic mapping:

$$x_1 = x_1' \cdot \frac{\theta}{\sin(\theta)}, \tag{7.14a}$$

$$x_2 = x_2' \cdot \frac{\theta}{\sin(\theta)}, \tag{7.14b}$$

where $\theta = \cos^{-1}(x_3')$. In case $\theta = 0$, $x_i = x_i'$ for $i = 1, 2$.

At the first step of the approximation algorithm, all the input points are mapped to the tangent plane, which has been proven as a linear space. These mapped points are denoted as $\mathbf{Q}_0, \ldots, \mathbf{Q}_m$. For those points on the tangent plane, we can apply the standard technique of linear least squares fitting [39].

To avoid the nonlinear problem, we pre-compute the parameters and knots. Then a linear least squares problem is set up and solved for the unknown control points. We choose the chord length to define the parameters. The distribution of the knots reflects the distribution of the parameters u_i. If $N = n + 1$ denotes the number of the control points for the dual spherical spline, p is the degree of the spline, we need a total of $N + p + 1$ knots. There are $N - p + 1$ internal knots and $N - p$ internal knot spans. Let $M = m + 1$ be the number of the given points:

$$d = \frac{M}{N - p}. \tag{7.15}$$

Then the internal knots are defined by:

$$\begin{aligned}
&i = int(jd); \\
&\alpha = jd - i; \\
&u_{p+j} = (1 - \alpha)u_{i-1} + \alpha u_i; \ j = 1, 2, \ldots, N - p + 1.
\end{aligned} \tag{7.16}$$

Here $i = int(jd)$ denotes the largest integer such that $i \leq jd$.

For the given m points $\mathbf{Q}_0, \ldots, \mathbf{Q}_m$ on the tangent plane, we seek a p-th degree B-spline $\mathbf{C}(u) = \sum_{i=1}^{n} N_i^p \mathbf{P}_i$, where N_i^p are p-th degree B-spline basis functions and $\mathbf{P}_i$ are its control points. This curve satisfies:

- $\mathbf{Q}_0 = \mathbf{C}_0$ and $\mathbf{Q}_m = \mathbf{C}_m$;
- the remaining $\mathbf{Q}_k$ are approximated in the least squares sense:

$$f = \sum_{k=1}^{m-1} |\mathbf{Q}_k - \mathbf{C}(u_k)|^2 \tag{7.17}$$

is a minimum with respect to the $n + 1$ variables. u_k are the pre-computed parameter values.

This approximated spline curve does not pass precisely through $\mathbf{Q}_i$ and $\mathbf{C}(u_i)$ is not the closest point on $\mathbf{C}(u)$ to $\mathbf{Q}_i$ [39]. Let

$$\mathbf{R}_k = \mathbf{Q}_k - N_{0,p}(u_k)\mathbf{Q}_0 - N_{n,p}(u_k)\mathbf{Q}_m, \ k = 1, \ldots, m - 1. \tag{7.18}$$

Then

$$
\begin{aligned}
f &= \sum_{k=1}^{m-1} |\mathbf{Q}_k - \mathbf{C}(u_k)|^2 \\
&= \sum_{k=1}^{m-1} |\mathbf{R}_k - \sum_{i=1}^{n-1} N_{i,p}(u_k)\mathbf{P}_i|^2 \\
&= \sum_{k=1}^{m-1} (\mathbf{R}_k - \sum_{i=1}^{n-1} N_{i,p}(u_k)\mathbf{P}_i) \cdot (\mathbf{R}_k - \sum_{i=1}^{n-1} N_{i,p}(u_k)\mathbf{P}_i) \\
&= \sum_{k=1}^{m-1} [\mathbf{R}_k \cdot \mathbf{R}_k - 2\sum_{i=1}^{n-1} N_{i,p}(u_k)(\mathbf{R}_k \cdot \mathbf{P}_i) \\
&\quad + (\sum_{i=1}^{n-1} N_{i,p}(u_k)\mathbf{P}_i) \cdot (\sum_{i=1}^{n-1} N_{i,p}(u_k)\mathbf{P}_i)].
\end{aligned}
\tag{7.19}
$$

To minimize f, we set the derivatives of f with respect to the $n-1$ control points $\mathbf{P}_1, \ldots, \mathbf{P}_{n-1}$ equal to zero:

$$
\frac{\partial f}{\partial \mathbf{P}_l} = 0,
\tag{7.20}
$$

which implies that:

$$
\sum_{i=1}^{n-1} (\sum_{k=1}^{m-1} N_{l,p}(u_k)N_{i,p}(u_k))\mathbf{P}_i = \sum_{k=1}^{m-1} N_{l,p}(u_k)\mathbf{R}_k.
\tag{7.21}
$$

Eq. (7.21) leads to a linear equation for the unknowns $\mathbf{P}_1, \ldots, \mathbf{P}_{n-1}$, it can be written as:

$$
(N^T N)\mathbf{P} = \mathbf{R},
\tag{7.22}
$$

where N is a $(m-1) \times (n-1)$ matrix of scalars:

$$
N = \begin{pmatrix}
N_{1,p}(u_1) & \ldots & N_{n-1,p}(u_1) \\
\vdots & \ddots & \vdots \\
N_{1,p}(u_{m-1}) & \ldots & N_{n-1,p}(u_{m-1})
\end{pmatrix}.
\tag{7.23}
$$

$\mathbf{R}$ is the vector of $n-1$ points:

$$
\mathbf{R} = \begin{pmatrix}
N_{1,p}(u_k)\mathbf{R}_1 + \ldots + N_{1,p}(u_{m-1})\mathbf{R}_{m-1} \\
\vdots \\
N_{n-1,p}(u_k)\mathbf{R}_1 + \ldots + N_{n-1,p}(u_{m-1})\mathbf{R}_{m-1}
\end{pmatrix},
\tag{7.24}
$$

and

$$
\mathbf{P} = \begin{pmatrix}
\mathbf{P}_1 \\
\vdots \\
\mathbf{P}_{n-1}
\end{pmatrix}.
\tag{7.25}
$$

This linear problem can be solved to get the control points $\mathbf{P}_1, \ldots, \mathbf{P}_{n-1}$ of the spline on the tangent plane. After deriving the control points, we need to apply the exponential mapping to get the control points $\mathbf{P}'_1, \ldots, \mathbf{P}'_{n-1}$ of the real spherical spline.

Now we use different inputs to verify the algorithm. Fig. 7.6 shows a sequence with 14 input points. Those points can be approximated very well by a Bézier curve, so we can see that the approximated spherical B-spline with 4 control points gives a good approximation result for the input points. Fig. 7.7 shows another sequence of input points. This figure gives more information about the approximation algorithm. We see that the approximated spherical B-spline does not pass precise every given points. The more control points we use, the better the approximation result is. Fig. 7.7(b), Fig. 7.7(c) and Fig. 7.7(d) show the approximated cubic B-spline with 4, 6 and 10 control points. Obviously, the approximation result gets improvement when more control points are adopted.

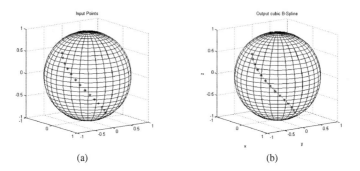

(a)   (b)

Fig. 7.6 Result I of the real spherical spline approximation algorithm: (a) Input for the algorithm; (b) Approximated spherical cubic B-spline with 4 control points.

7.4.2 Dual spherical spline approximation algorithm

After deriving the real spherical spline approximation algorithm, we can develop the dual spherical spline approximation algorithm similarly.

The dual spherical spline approximation is developed based on a mapping between the DUS and a tangent plane. The tangent plane is chosen corresponding to the middle point of the given input point sequence. This point should be mapped to $(0,0,1)$ by rotating the coordinate system using Eq. (5.2). Then, the points on the this tangent plane is mapped to the DUS by the exponential mapping defined in Eq. (4.20) and Eq. (4.21). Reversely, the points on the DUS are mapped to the tangent plane by the logarithmic mapping defined in Eq. (4.22). Those steps are similar as the real case, only different calculation rules are applied. After the mapping, a sequence of points on the tangent plane are derived, denoted as $\hat{\mathbf{Q}}_0, \ldots, \hat{\mathbf{Q}}_m$. Then the standard technique of linear least squares fitting is applied[39].

First all. we need to determine the parameters and knot sequence. In our program, the chord length is chosen to define the parameters: let $\hat{d}_i$ be the chord length between two given points: $\hat{d}_i = |\hat{\mathbf{Q}}_i - \hat{\mathbf{Q}}_{i-1}|, i = 1, \ldots, m$, the total chord length is calculated as $\hat{d} = \sum_{i=1}^{m} \hat{d}_i$. Since $\hat{d}_i$ is a dual number, we take the principle part of $\hat{d}_i$ as d_i, the parameters are calculated as:

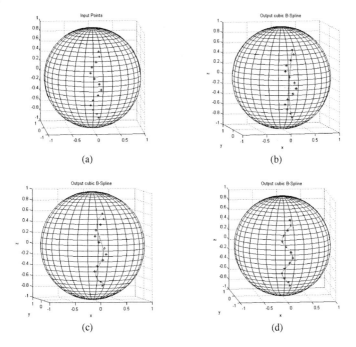

Fig. 7.7 Result II of the real spherical spline approximation algorithm: (a) Input for the algorithm; (b) Approximated spherical cubic B-spline with 4 control points; (c) Approximated spherical cubic B-spline with 6 control points; (d) Approximated spherical cubic B-spline with 10 control points.

$$u_0 = 0;$$
$$u_i = u_{i-1} + \frac{d_i}{d}, \ k = 1, \dots, m-1,$$
$$u_m = 1.$$
(7.26)

After deriving the parameter, the same technique described in previous section is adopted to determine the knots sequence.

Now, for the given m points $\hat{\mathbf{Q}}_0, \dots, \hat{\mathbf{Q}}_m$ on the tangent plane, we seek a p-th degree dual B-spline $\hat{\mathbf{C}}(u) = \sum_{i=1}^{n} N_i^p \hat{\mathbf{P}}_i$, where N_i^p are p-th degree B-spline basis functions and $\hat{\mathbf{P}}_i$ are its control points. This curve satisfies:

- $\hat{\mathbf{Q}}_0 = \hat{\mathbf{C}}_0$ and $\hat{\mathbf{Q}}_m = \hat{\mathbf{C}}_m$;
- the remaining $\hat{\mathbf{Q}}_k$ are approximated in the least squares sense:

$$\hat{f} = \sum_{k=1}^{m-1} |\hat{\mathbf{Q}}_k - \hat{\mathbf{C}}(u_k)|^2$$
(7.27)

is a minimum with respect to the $n+1$ variables. u_k are the pre-computed parameter values.

The following operation is almost the same as the real case, only the calculation is based on the calculation rules defined in dual vector space. Let

$$\hat{\mathbf{R}}_k = \hat{\mathbf{Q}}_k - N_{0,p}(u_k)\hat{\mathbf{Q}}_0 - N_{n,p}(u_k)\hat{\mathbf{Q}}_m, \quad k = 1,\ldots,m-1. \tag{7.28}$$

Then

$$\begin{aligned}
\hat{f} &= \sum_{k=1}^{m-1} |\hat{\mathbf{Q}}_k - \hat{\mathbf{C}}(u_k)|^2 \\
&= \sum_{k=1}^{m-1} |\hat{\mathbf{R}}_k - \sum_{i=1}^{n-1} N_{i,p}(u_k)\hat{\mathbf{P}}_i|^2 \\
&= \sum_{k=1}^{m-1} (\hat{\mathbf{R}}_k - \sum_{i=1}^{n-1} N_{i,p}(u_k)\hat{\mathbf{P}}_i) \cdot (\hat{\mathbf{R}}_k - \sum_{i=1}^{n-1} N_{i,p}(u_k)\hat{\mathbf{P}}_i) \\
&= \sum_{k=1}^{m-1} [\hat{\mathbf{R}}_k \cdot \hat{\mathbf{R}}_k - 2\sum_{i=1}^{n-1} N_{i,p}(u_k)(\hat{\mathbf{R}}_k \cdot \hat{\mathbf{P}}_i) \\
&\quad + (\sum_{i=1}^{n-1} N_{i,p}(u_k)\hat{\mathbf{P}}_i) \cdot (\sum_{i=1}^{n-1} N_{i,p}(u_k)\hat{\mathbf{P}}_i)].
\end{aligned} \tag{7.29}$$

Now $\hat{f}$ is a dual-numbered function of $n-1$ dual vectors $\hat{\mathbf{P}}_1,\ldots,\hat{\mathbf{P}}_{n-1}$. To minimize $\hat{f}$, we set the derivatives of $\hat{f}$ with respect to the $n-1$ control points $\hat{\mathbf{P}}_1,\ldots,\hat{\mathbf{P}}_{n-1}$ equals to zero:

$$\frac{\partial \hat{f}}{\partial \hat{\mathbf{P}}_l} = 0, \tag{7.30}$$

which implies that:

$$\sum_{i=1}^{n-1} (\sum_{k=1}^{m-1} N_{l,p}(u_k)N_{i,p}(u_k))\hat{\mathbf{P}}_i = \sum_{k=1}^{m-1} N_{l,p}(u_k)\hat{\mathbf{R}}_k. \tag{7.31}$$

It can be written as:

$$(N^T N)\hat{\mathbf{P}} = \hat{\mathbf{R}}, \tag{7.32}$$

where N is a $(m-1) \times (n-1)$ matrix of scalars:

$$N = \begin{pmatrix} N_{1,p}(u_1) & \ldots & N_{n-1,p}(u_1) \\ \vdots & \ddots & \vdots \\ N_{1,p}(u_{m-1}) & \ldots & N_{n-1,p}(u_{m-1}) \end{pmatrix}. \tag{7.33}$$

$\hat{\mathbf{R}}$ is the vector of $n-1$ points:

$$\hat{\mathbf{R}} = \begin{pmatrix} N_{1,p}(u_k)\hat{\mathbf{R}}_1 + \ldots + N_{1,p}(u_{m-1})\hat{\mathbf{R}}_{m-1} \\ \vdots \\ N_{n-1,p}(u_k)\hat{\mathbf{R}}_1 + \ldots + N_{n-1,p}(u_{m-1})\hat{\mathbf{R}}_{m-1} \end{pmatrix}, \tag{7.34}$$

and

$$\hat{\mathbf{P}} = \begin{pmatrix} \hat{\mathbf{P}}_1 \\ \vdots \\ \hat{\mathbf{P}}_{n-1} \end{pmatrix}. \tag{7.35}$$

This linear problem can be solved to get the control points $\hat{\mathbf{P}}_1,\ldots,\hat{\mathbf{P}}_{n-1}$ of the dual spline on the tangent plane. After deriving the control points, we need to apply the exponential mapping to get the control points $\hat{\mathbf{P}}'_1,\ldots,\hat{\mathbf{P}}'_{n-1}$ of the dual spherical spline on the DUS. Finally, the dual spherical spline is evaluated based on the algorithm of calculating the weighted average on the DUS.

7.4.3 Simulation and application in optimization

We use the same blade file used earlier in this chapter as a test example of the approximation algorithm and reduce the number of the control points gradually. The results are shown in Fig. 7.8. Obviously, the resulting ruled surface does not pass all the input rulings, this is a natural result that the approximated spline does not require to pass precisely through every given point. We also notice that, in the distance sense, the approximation result improves as the number of control points increases. However, as the number of control points approaches the number of data points, undesirable shapes may occur if the data contains noise.

Applying this dual spherical spline approximation algorithm instead of the interpolation algorithm, we can control the number of the parameters for the optimizer in the initialization stage. Fig. 7.9 shows the results of the initialization step, using 10, 7 and 4 control points respectively. We can see that the constructed blades still keep similar shape as the given blade, however, the less the control points are used, the worse the approximation becomes. Fig. 7.10 shows the comparison between the given blade with the constructed blades using different number of control points. For the last case, which only uses 4 control points, the constructed blade is a little bit deformed. But in the remaining two cases, the initialization results are still acceptable. Table 7.3, Table 7.4 and Table 7.5 show the control points mapped to the tangent plane, from which we can see that the first and the last control points remain unchanged. Those control points are the parameters for the optimizer.

Control Point 1	$(-0.1965, -0.8383, 0) + \varepsilon(-17.5712, 2.3586, 0)$
Control Point 2	$(-0.1502, -0.8043, 0) + \varepsilon(-16.4270, 1.7151, 0)$
Control Point 3	$(-0.0399, -0.6571, 0) + \varepsilon(-14.4428, 0.1285, 0)$
Control Point 4	$(0.0757, -0.3855, 0) + \varepsilon(-8.1692, -1.3932, 0)$
Control Point 5	$(0.0621, 0.0218, 0) + \varepsilon(-0.0885, -1.0028, 0)$
Control Point 6	$(-0.2511, 0.2562, 0) + \varepsilon(7.4076, 4.2150, 0)$
Control Point 7	$(-0.5905, 0.0688, 0) + \varepsilon(9.6669, 9.5687, 0)$
Control Point 8	$(-0.7995, -0.0200, 0) + \varepsilon(10.7780, 12.7591, 0)$
Control Point 9	$(-0.9018, -0.2759, 0) + \varepsilon(9.8718, 14.8640, 0)$
Control Point 10	$(-0.8696, -0.3096, 0) + \varepsilon(8.8823, 14.6783, 0)$

Table 7.3 Ten control points of the dual spherical spline on the tangent plane.

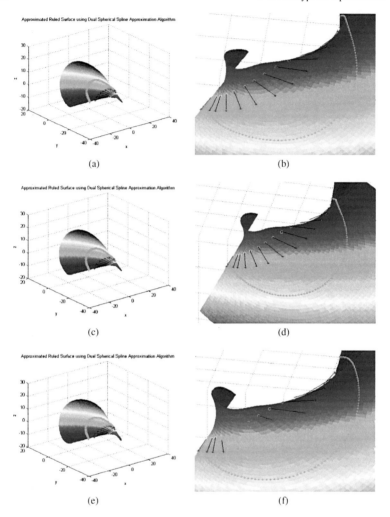

Fig. 7.8 Result of the dual spherical spline approximation algorithm: (a) Approximation result using 10 control points; (b) Details for approximation result using 10 control points; (c) Approximation result using 7 control points; (d) Details for approximation result using 7 control points; (e) Approximation result using 4 control points; (f) Details for approximation result using 4 control points.

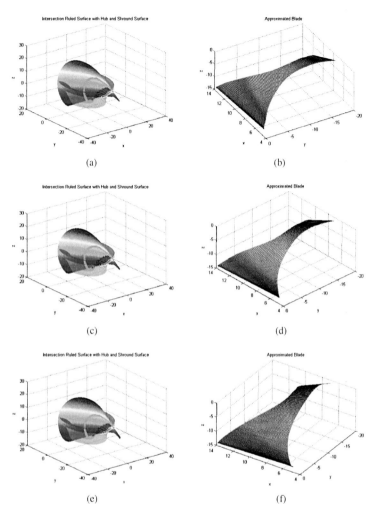

Fig. 7.9 Application of the dual spherical spline approximation algorithm in the optimizer initialization: (a) Intersecting the approximated ruled surface (10 control points) with hub and shroud surfaces; (b) Constructed blade with 10 control points; (c) Intersecting the approximated ruled surface (7 control points) with hub and shroud surfaces; (d) Constructed blade with 7 control points; (e): Intersecting the approximated ruled surface (4 control points) with hub and shroud surfaces; (f) Constructed blade with 4 control points.

7.5 Comments

In this section, we propose an initial prototype for the ruled surface optimization. The dual spherical spline interpolation algorithm is extended to the dual spheri-

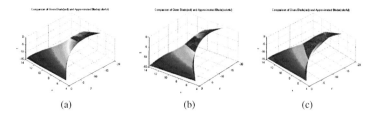

(a)   (b)   (c)

Fig. 7.10 Comparison between the constructed blade and the given blade: (a) 10 control points; (b) 7 control points; (c) 4 control points.

Control Point 1	$(-0.1965, -0.8383, 0) + \varepsilon(-17.5712, 2.3586, 0)$
Control Point 2	$(-0.0532, -0.6991, 0) + \varepsilon(-14.6174, 0.3834, 0)$
Control Point 3	$(0.2090, -0.2286, 0) + \varepsilon(-5.9015, -3.4790, 0)$
Control Point 4	$(-0.2330, 0.3847, 0) + \varepsilon(9.2821, 4.0191, 0)$
Control Point 5	$(-0.8020, -0.0644, 0) + \varepsilon(10.2849, 12.7900, 0)$
Control Point 6	$(-0.8572, -0.1469, 0) + \varepsilon(10.5584, 13.9803, 0)$
Control Point 7	$(-0.8696, -0.3096, 0) + \varepsilon(8.8823, 14.6783, 0)$

Table 7.4 Seven control points of the dual spherical spline on the tangent plane.

Control Point 1	$(-0.1965, -0.8383, 0) + \varepsilon(-17.5712, 2.3586, 0)$
Control Point 2	$(0.5917, -0.2360, 0) + \varepsilon(-7.9434, -8.8625, 0)$
Control Point 3	$(-0.5776, 0.7431, 0) + \varepsilon(17.4594, 8.5640, 0)$
Control Point 4	$(-0.8696, -0.3096, 0) + \varepsilon(8.8823, 14.6783, 0)$

Table 7.5 Four control points of the dual spherical spline on the tangent plane.

cal spline approximation algorithm. The two directrix curves of a ruled surface are determined by an intersection algorithm which calculates the intersection curve between a ruled surface and a free-form surface. Employing the exponential and the logarithmic mapping, we reduce one third number of parameters for the optimizer. In this dissertation, we implement the initialization phase for this optimization model, the constructed ruled surface and its parameters are shown. This prototype can be integrated in certain optimizer to derive an optimal ruled surface for a blade design. Since this optimal ruled surface is defined as a dual spherical spline, which describes the movement of a line which generates the ruled surface, it has direct link to the manufacturing process. It can be used to guide the movement of a manufacturing tool, which is regarded as a line, such as wire EDM and laser cutting. However, for more general cases, the milling tool has certain size and shape, for example, the cylindrical cutter. Therefore, we need to take the shape of the milling tool into consideration. In this case, the ideal position for the cutting tool is to offset the ruling in the direction of a surface normal at a distance that equals the radius of the cutting tool. We are going to discuss this application in Chapter 8.

Chapter 8
Application in CNC Machining

In this chapter, we extend the new developed algorithm to the applications in CNC machining. As we know, a curved surface can be manufactured by point milling method, but the manufacturing cost is high. To reduce the manufacturing cost, the curved surface is approximated with a ruled surface and manufactured by flank milling method, which brings in approximation error. Theoretically, if the machine tool is considered as a line, a ruled surface can be accurately produced by moving this line. But in real life the machine tool usually has certain size and shape (i.e., cylindrical cutter or conical cutter), so the ideal position for the cutting tool is to offset the ruling in the direction of a surface normal at a distance equal to the radius of the cutting tool. Because the surface normals rotate along the ruling, at some point, the cutting tool will begin to deviate from the desired surface. Generally, the machined surface is not a ruled surface, but a curved surface. At each tool position, the effective contact between the cutting tool and the swept surface is a curve (grazing curve), not a straight line. Hence, we propose a new strategy to design a desired surface by specifying the tool path that generates it. Fig. 8.1 compares this new approach with the conventional design and manufacturing methods. The new approach provides a novel design and manufacturing strategy which ensures low manufacturing cost and avoids introducing double error.

The first part of the chapter describes the 5-axis flank milling procedure in CNC machining. Next, an algorithm which generates a tool path for the flank milling method is proposed. It mainly contains two parts: the offset surface generation and drive surface construction. Specifically, the path generation is presented for cylindrical milling tool. Different from the traditional methods, we skip the calculation of cutter-contact paths, but directly focus on the construction of the drive surface corresponding to the axis of cylindrical cutter (CL data). Obviously, the drive surface which is generated by the displacement of the center line of cylindrical cutter is a ruled surface. It can be represented by the new defined dual spherical spline. Therefore, we can combine the design and manufacturing in one step by applying a ruled surface approximation algorithm to obtain the drive surface. This algorithm is tested by some given turbocharger blades. The simulation results are presented.

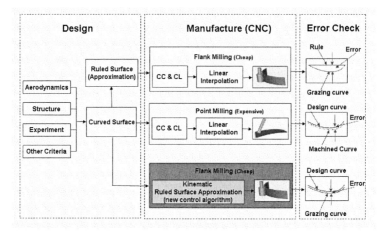

Fig. 8.1 A comparison of design and manufacturing diagram.

This algorithm can also be generalized to the path generation of face milling in CNC machining.

8.1 5-axis Flank Milling in CNC Machining

Computer Numerical Control (CNC) machining is a widely used machining technique in today's manufacturing industry. The machines are automatically operated by commands received by their processing units. The commands are generated by a Computer Aided Manufacturing (CAM) system and transferred to the CNC unit to control the motion of the machine tool. With the help of computer, the machining process is more accurate and efficient.

8.1.1 5-axis machine

Nowadays, 5-axis CNC machining is becoming the main stream due to its ability to handle geometrically complex workpieces composed of different raw material. A typical 5-axis milling machine carries three translational and two rotary axes. Figure 8.2 shows one example of such a CNC machine. The five degrees of freedom provide the minimum need for establishing an arbitrary position and orientation of the cutting tool relative to the workpiece. Three spatial degrees are used to locate the tool at the cutter location point; the extra two rotational degrees are used to establish the orientation of the tool represented by the *inclination angle* and *tilt angle*.

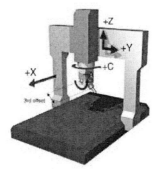

Fig. 8.2 A CNC machine with 2 rotational axes and 3 translational axes.

Besides the most common type of 5-axis milling machine, there are three other groups due to the different number of translational and rotary axes: (i) two translational axes and three rotary axes; (ii) one translational axes and four rotary axes; (iii) five rotary axes. The machine axes can be assigned to the tool and to the machine table. Based on the above classification criteria, the 5-axis machines can be divided into six groups [25]:

1. 5/0 machine: The table carries all axes and the tool is fixed in space;
2. 4/1 machine: The table carries four axes and the tool carries one axis;
3. 3/2 machine: The table carries three axes and the tool carries two axes;
4. 2/3 machine: The table carries two axes and the tool carries three axes;
5. 1/4 machine: The table carries one axis and the tool carries four axes;
6. 0/5 machine: The tool carries all axes and the table is fixed in space.

Figure 8.3 shows a draft of a 2/3 machine. A block of a raw material called *workpiece* is fixed to the *machine table* which carries two axes. The material is removed from the workpiece by a rotating cutter attached to the *spindle* through the *tool holder*, which carries three axes. The process of material removal with the goal to produce a required industrial part is called *milling*, *machining* or *cutting*.

The CNC machines are programmed by means of a special code called the *NC program*. The program consists of instructions to control the machine movements following a certain manufacturing technology. The generation of NC program in 5-axis CNC machining requires four steps [7] [55]:

- Tool path generation in the workpiece coordinate system: This step requires a successive set of coordinates called *Cutter Contact* (CC) points $W = (x_w, y_w, z_w)$ and the tool orientations $I = (I_x, I_y, I_z)$. The CC points are the set of points where the edge of the tool passes through. The cutter contact points are derived from the input surface data, each CC point is associated with a tool vector positioned in a right-handed, rectangular coordinate system.

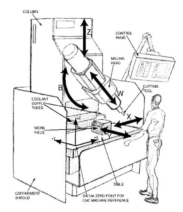

Fig. 8.3 A draft of 5-axis CNC machine.

- *Cutter Location* (CL) data obtain: The CL data are obtained from the CC data. The conversion from CC data to CL data depends on the tool geometry and inclination. Typically, the tool trajectory is generated by determining the CC points and offsetting them to generate the CL points. Then the CL data are converted to joint values of the 5-axis machine.
- Bounded-deviation joint path planning: In this step, an interpolation technique is adopted to convert the point data into the motion of the tool axes. A simple way is to employ linear interpolation techniques, which breaks down the tool path into a set of straight line segments. It leads to a command path consisting of linear motions between points. However, the simple interpolation methods may result in discontinuities of the velocity at the junctions of the segments. Sometimes, they may also lead to high accelerations or subsequent surface inaccuracies that cost much time to eliminate. Hence, many modern interpolation schemes are focused on finding a suitable parametric interpolation. A parametric interpolation takes the tool path in a parametric form and the command path is a function of the curve parametric and the desired feed-rate.
- *Post-process*: The tool path is transformed to the machine coordinate system and the NC code is obtained. This process requires information corresponding to a particular 5-axis machine.

Due to the variety of 5-axis CNC machine types, we exclude the post-process step in this dissertation, but focus on finding an appropriate parametric interpolation technique which provides a smooth tool path for the given designed surface in the coordinate system of workpiece.

Fig. 8.4 Manufacture an impeller with a 5-axis CNC machine.

8.1.2 Flank milling

The 5-axis CNC machining of sculptured surface can be classified into two types: *face milling* and *flank milling*. Face milling is also called *point milling*, which uses the tip of the milling tool (e.g., flat-end cutter). Different from the face milling method, flank milling (*side milling*) uses the side of the manufacturing tool instead of the tip of the manufacturing tool touches the surface and removes the stock in front of the cutter. Since the whole length of the cutter is involved in the cutting process, this method has high material removal rate and high machining efficiency. Besides, no scallops are left behind in single pass flank milling, less surface finishing work is required [30]. The advantages of flank milling attract many engineers and researchers.

In industry, the 5-axis flank milling is often used to machine ruled surfaces. It is widely used to machine turbine blades, fan impellers and other products. In general, the face milling with flat-end cutter is suitable for machining large sculptured surface, while the flank milling with cylindrical cutter has more applications milling small and middle dimensional surfaces. However, the manufacturing of a turbocharger compressor/impeller is necessary to use 5-axis flank milling, because the tunnel between two adjacent blades is too small with respect to the size of blades. Fig. 8.4 shows an example of flank milling an impeller with a cylindrical cutter in 5-axis CNC machine.

In flank milling, a convectional way of generating the CL data to mill a ruled surface is to locate the center line (axis) of the tool with respect to rulings of the surface. For cylindrical cutters, the CL data can be derived by relocating the ruling in the direction of the surface normals by a distance equal to the radius of the cylinder. However, this method only works for developable surfaces. For non-developable ruled surface, it is impossible to position a cylindrical cutter without either *under-cutting* or *overcutting* the desired surface [27]. Overcutting (gouging) means that the cutting tool will cut into the machined surface; and undercutting means that

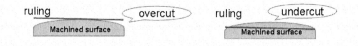

Fig. 8.5 Two types of manufacturing error.

the cutting tool will depart from the machined surface. Figure 8.5 shows the two different types of manufacturing error.

In order to minimize the manufacturing error, researchers developed various CL data optimization methods. The simplest way is to locate the cylindrical cutting tool tangentially to the given surface at one point on the ruling and make the tool axis parallel to the ruling. Alternatively, the tool can also be positioned to touch two points on the ruling. Both ideas belong to the direct tool position method [24]. An improvement of the direct tool position method is to locate the tool step by step [7] [48] [30]. In those approaches, the initial position of the cutting tool is determined by one of the direct tool position methods, afterwards the tool is lifted and twisted in order to reduce the manufacturing error. The computation time of the step by step method is usually long. The third type of tool positioning method combines the techniques used in the two classes above. The tool contacts three points on the given surface (two on the guiding curves and one on the ruling). Those three points are obtained by solving seven transcendental equations based on certain geometrical conditions [47]. Fig. 8.6 shows an optimization process which generates a program of flank milling in CNC machining.

However, those methods all focus on the local error reduction respect to each tool location. The kinematic error between successive CL points can still be large. In order to get a global optimal tool path, a new type of approach is developed [18] [8] [50] [65]. The authors propose a global optimization method to generate the tool axis trajectory surface (which is also a ruled surface). The cutting tool is positioned so that the maximum deviation between the tool axis trajectory surface and the offset surface is minimized. They first offset the given surface and generate a serious of sample points, then adopt the least squares surface fitting method to find a B-spline surface that closely matches the sample points of the offset surface. This B-spline surface is the drive surface which is the trajectory surface of the tool axis, therefore each tool position is determined.

All those methods described above attempt to reduce the deviation between the machined surface and the designed surface. For the turbocharger design, the separated design and manufacturing phases bring in error twice: The designers make a lot of effort to improve the efficiency and performance of the turbocharger. In order to reduce the manufacturing cost, the original design is approximated with a ruled surface. However, using flank milling method with cylindrical cutter in 5-axis

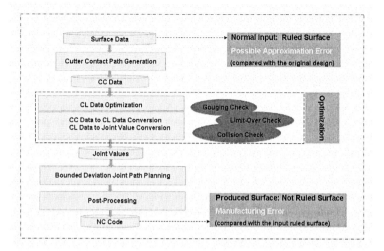

Fig. 8.6 A conventional flow of flank milling in 5-axis CNC machining.

CNC machining, the machined surface is not a ruled surface but a curved surface. Recently, researchers start to analyze the mathematical expression of the machined surface and use a swept surface to approximate the machined surface [23]. This is a fresh perspective on combining design and manufacturing. However, it does not provide a direct expression of the tool path.

Inheriting the spirit of our new developed algorithm, we propose a tool planing approach for flank milling with cylindrical cutter in 5-axis machining. This strategy combines the design and manufacturing stages together, providing a tool path in the form of a dual spherical spline. As we know, an effective method for specifying a 5-axis CNC machining tool path is to relate both the position and orientation to a single parameter. Due to the compact data structure, a single parameter smooth dual spherical spline is derived as a tool path. Each point on the spline corresponds to a line which is represented by a dual vector indicating the orientation of the tool axis with respect to a specific point on the workpiece. The point on the tool can be chosen from either the contact point or the tool center point. In this way, the path of a 5-axis CNC machining tool is uniquely specified. Using the definition of the screw and the dual vector calculation rules, the tool axis position is easily converted to the tool motion. Besides, the dual spherical spline represents the operation of the CNC machine accurately. Applying the kinematics and robotics analysis, it is possible to check whether the desired path is within the workspace of the machine tool. This novel tool path representation also offers some convenience in determining manufacturing error.

8.2 Path Generation

As stated before, the tool path for the 5-axis CNC machining is uniquely specified by the path of an arbitrary point on the tool axis and a vector indicating the direction along the axis. An effective way to represent the tool path is to link both the position and orientation to a single parameter. The orientation and the translation must be expressed in terms of a smooth motion across the workpiece [55]. The displacement of the tool axis of a cylindrical cutter generates a ruled surface. In other words, the drive surface is a ruled surface. This discovery leads us to a practical method of representing the tool path of 5-axis CNC machining. The essence of this method contains two parts: first, the offset of the given surface is derived; second, the offset surface is approximated with a ruled surface. Adopting the kinematic ruled surface approximation algorithm, this ruled surface is derived as a dual spherical spline.

8.3 Offset Theory

In this section, a brief introduction of offset theory is presented. Different types of offset definitions are introduced. Those concepts are very important to understand the principle of the path planning in CNC machining and useful for algorithm development.

If $\mathbf{R}(\mathbf{u}) = \mathbf{R}(u_1, u_2)$ represents a surface, its *offset surface* $\mathbf{R}_o(\mathbf{u})$ is defined by the equation [27]:

$$\mathbf{R}_o(\mathbf{u}) = \mathbf{R}(\mathbf{u}) + d \cdot \mathbf{n}(\mathbf{u}), \tag{8.1}$$

where $\mathbf{n}$ is a normal vector in $\mathbf{R}(\mathbf{u})$ and d is the distance between the surfaces. This is the classical definition of offset surface. It is also referred to as *parallel offset*. In [43], H. Pottmann and W. Lü study the "*circular offset*" of ruled surfaces, which arise when a cylindrical or conical cutter with a circular edge is used in flank milling. The authors proved that the circular offsets of a rational ruled surface are rational in general except the developable surfaces and conoidal ruled surfaces with generators orthogonal to the tool-axis. However, the offset of a ruled surface is in general not a ruled surface. In fact, the offset curve of a nontorsal generator with respect to a ruled surface is a rational quadric [43].

For ruled surface, we sometimes use the concept of *Bertrand offset*. It is a generalization of the theory of *Bertrand curves* based on line geometry. A pair of curves are Bertrand mates if there exists a one-to-one correspondence between their points such that both curves share a common principal normal at their corresponding points [45]. According to this definition, given a space curve $\mathbf{r}(u)$ and its Bertrand offset $\mathbf{r}'(u)$, the principal normal $\mathbf{n}$ of $\mathbf{r}(u)$ is the same as the principal normal $\mathbf{n}'$ of $\mathbf{r}'(u)$. The following equation is derived:

$$\mathbf{r}'(u) = \mathbf{r}(u) + R\mathbf{n}(u), \tag{8.2}$$

where R is the offset distance between the two corresponding points, and the definition of the principal normal $\mathbf{n}$ is consistent with the Frenet trihedron of the curve:

$$\begin{cases} \mathbf{t} = \mathbf{r}_s = \frac{\mathbf{r}_u}{|\mathbf{r}_u|}, \\ \mathbf{n} = \frac{\mathbf{r}_{ss}}{|\mathbf{r}_{ss}|} = \frac{(\mathbf{r}_u \times \mathbf{r}_{uu}) \times \mathbf{r}_u}{|(\mathbf{r}_u \times \mathbf{r}_{uu})| \cdot |\mathbf{r}_u|}, \\ \mathbf{b} = \frac{\mathbf{r}_s \times \mathbf{r}_{ss}}{|\mathbf{r}_{ss}|} = \frac{\mathbf{r}_u \times \mathbf{r}_{uu}}{|\mathbf{r}_u \times \mathbf{r}_{uu}|}. \end{cases} \tag{8.3}$$

where $s = \int_a^b |\mathbf{r}_u(u)| du$ is the arc length of $\mathbf{r}(u)$. It can be proven that the tangents to the corresponding points of a curve and its Bertrand offset make a constant angle θ. Then we derive the relationship between the Frenet trihedron of a curve and that of its Bertrand offset [45]:

$$\begin{pmatrix} \mathbf{t}' \\ \mathbf{n}' \\ \mathbf{b}' \end{pmatrix} = \begin{pmatrix} \cos\theta & 0 & \sin\theta \\ 0 & 1 & 0 \\ -\sin\theta & 0 & \cos\theta \end{pmatrix} \begin{pmatrix} \mathbf{t} \\ \mathbf{n} \\ \mathbf{b} \end{pmatrix}. \tag{8.4}$$

Considering the ruled surface in the context of line geometry, the ruled surface is represented as a one-parameter family of lines. The theory analogous to the theory of Bertrand curves is developed. In line geometry, the offset distance between two lines is defined in terms of a linear and an angular offset. The linear offset R is the length of the common perpendicular between the two lines. The angular offset θ is the angle between the two lines measured in a plane orthogonal to the common perpendicular between the two lines. Simply speaking, we have the following definition [45]:

definition 8.3.1 *Two ruled surfaces are said to be Bertrand offsets of one another if there exists a one-to-one correspondence between their rulings such that both surfaces have a common principal normal at the striction points of their corresponding rulings.*

If a ruled surface $\mathbf{L}(s,t)$ is given as:

$$\mathbf{L}(s,t) = \mathbf{c}(s) + t\mathbf{e}(s), \tag{8.5}$$

where $\mathbf{c}$ is the striction curve, $\mathbf{e}$ is the spherical indicatrix and s is the arc length along $\mathbf{c}$. The Frenet trihedron on the ruled surface is defined by the vectors $\mathbf{e}, \mathbf{t}$ and $\mathbf{g}$. $\mathbf{t}$ is the central normal and $\mathbf{g}$ is the asymptotic normal, which are defined by the following equations [45]:

$$\begin{aligned} \mathbf{t} &= \frac{\mathbf{e}_s}{|\mathbf{e}_s|}, \\ \mathbf{g} &= \frac{\mathbf{e} \times \mathbf{e}_s}{|\mathbf{e}_s|}. \end{aligned} \tag{8.6}$$

Under the above definition, we can get a similar relationship as Eq. (8.4) for the geodesic Frenet trihedron of the based ruled surface and its Bertrand offset:

$$\begin{pmatrix} \mathbf{e}' \\ \mathbf{t}' \\ \mathbf{g}' \end{pmatrix} = \begin{pmatrix} \cos\theta & 0 & \sin\theta \\ 0 & 1 & 0 \\ -\sin\theta & 0 & \cos\theta \end{pmatrix} \begin{pmatrix} \mathbf{e} \\ \mathbf{t} \\ \mathbf{g} \end{pmatrix}. \tag{8.7}$$

Then, the offset surface $\mathbf{L}'$ is written as:

$$
\begin{aligned}
\mathbf{L}'(s,t) &= \mathbf{c}' + t\mathbf{e}', \\
&= [\mathbf{c}(s) + R\mathbf{t}(s)] + t[(\cos\theta)\mathbf{e}(s) + (\sin\theta)\mathbf{g}(s)],
\end{aligned} \tag{8.8}
$$

where R is the linear offset, θ is the angular offset. If $\theta = 0$, the corresponding rulings on two surfaces are parallel. These two surfaces are called *"oriented offsets"*. If $\theta = \pi/2$, the corresponding rulings on the two surfaces are perpendicular. These two surfaces are called *"right offsets"*. For the Bertrand offsets, we have an important theorem [45]:

theorem 8.3.1 *Two ruled surfaces which are Bertrand mates as define in Definition 8.3.1 are constant offsets of one another.*

Inspired by this theorem, if the given surface is a ruled surface[1], the drive surface can be derived by constructing the Bertrand offset of the given surface. Consequently, the given surface is also a Bertrand offset surface of the drive surface. This relationship provides the initial inspiration of our approach. However, it should be pointed out that oriented Bertrand offsets of a ruled surface are, in general, not parallel offsets. For an oriented Bertrand offset to become a parallel offset, the normals at different points of a pair of corresponding rulings should all be the same. This means that both the base surface and its Bertrand offset have to be developable [45].

Generally, the original designed surface is not a ruled surface. In our algorithm, we calculate the "circular offset" of the given surface instead of the "Bertrand offset". Because of this difference, the approximation error may be large when the given surface is designed far from the ruled surface. However, for our specific applications, we do not meet serious problems, since the size of the blade is small and the original design is usually close to a ruled surface.

8.3.1 Path generation and motion conversion

As long as the offset surface is derived, a kinematic ruled surface approximation algorithm can be applied to get a ruled surface. This ruled surface is the drive surface. The essence of this part is presented in Chapter 6, we are not going to repeat it. One boundary of the ruled surface is determined by intersecting a reference line with the center line. The other boundary is determined by the length of the cylindrical cutter.

Now the drive surface which is a ruled surface represented as a continuous, differentiable curve $\hat{\mathbf{x}}(u)$ on the dual unit sphere. Following this definition, a local coordinate frame can be set up consisting of three concurrent lines $\{\hat{\mathbf{x}}, \hat{\mathbf{n}}, \hat{\mathbf{t}}\}$. This frame is called *generator trihedron*, where $\hat{\mathbf{x}}$ represents a ruling and the other two lines are defined by the following equations:

---

[1] for the traditional design method, the surface is approximated as a ruled surface before the manufacturing.

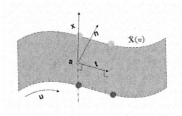

Fig. 8.7 Generator trihedron on a ruled surface.

$$\hat{\mathbf{t}} = \frac{\frac{d\hat{\mathbf{x}}(u)}{du}}{\left\| \frac{d\hat{\mathbf{x}}(u)}{du} \right\|},$$

$$\hat{\mathbf{n}} = \hat{\mathbf{x}} \times \hat{\mathbf{t}}. \tag{8.9}$$

The line $\hat{\mathbf{t}}$ is called the *central tangent*, which is tangent to the surface at the striction point. The line $\hat{\mathbf{n}}$ called *central normal* is the normal of the surface at the striction point. It can be proven that these three lines are orthogonal to each other and the intersection point of the three lines is the *striction point* of the ruling $\hat{\mathbf{x}}$. This point is the point of minimum distance between neighboring rulings. The locus of striction points is called the *striction curve* [54]. The generator trihedron can be rewritten as dual vectors:

$$\hat{\mathbf{x}} = \mathbf{x} + \varepsilon(\mathbf{a} \times \mathbf{x}),$$
$$\hat{\mathbf{n}} = \mathbf{n} + \varepsilon(\mathbf{a} \times \mathbf{n}), \tag{8.10}$$
$$\hat{\mathbf{t}} = \mathbf{t} + \varepsilon(\mathbf{a} \times \mathbf{t}),$$

where $\mathbf{a}$ is the striction point, $\mathbf{x}$ is a vector directing along the ruling, the vector $\mathbf{t}$ is perpendicular to $\mathbf{x}$ and tangent to the surface at the striction curve, the vector $\mathbf{n}$ is perpendicular to $\mathbf{x}$ and $\mathbf{t}$. Fig. 8.7 shows this frame on a ruled surface. The center line of the cylindrical cutter (a ruling of drive surface) undergoes a screw motion about the axis $\hat{\mathbf{t}}$. According to the screw theory, the distance between successive rulings is defined as a dual angle between two screws. The successive rulings are denoted as $\hat{\mathbf{x}}_1 = \hat{\mathbf{x}}(u_1) = \sum_{i=0}^{n} B_{i,p}(u_1)\hat{\mathbf{P}}_i$ and $\hat{\mathbf{x}}_2 = \hat{\mathbf{x}}(u_2) = \sum_{i=0}^{n} B_{i,p}(u_2)\hat{\mathbf{P}}_i$. The dual angle is calculated by the following equations:

$$\hat{\mathbf{x}}_1 \cdot \hat{\mathbf{x}}_2 = \cos\hat{\omega} = x + \varepsilon x^\circ, \tag{8.11a}$$

$$\begin{aligned}\hat{\omega} &= \phi + \varepsilon d \\ &= \cos^{-1}(\hat{x}) \\ &= \cos^{-1}(x + \varepsilon x^\circ) \\ &= \cos^{-1}(x) - \varepsilon\frac{x^\circ}{\sqrt{1-x^2}}.\end{aligned} \tag{8.11b}$$

This means the cutter tool translate distance d and rotate angle ϕ along the axis $\hat{\mathbf{t}}$ in order to move from position $\hat{\mathbf{x}}_1$ to position $\hat{\mathbf{x}}_2$. The ratio between d and ϕ is called *distribution parameter* [55]:

$$p = \frac{d}{\phi}. \tag{8.12}$$

The distribution parameter indicates the amount of twisting associated with the ruled surface. A cone or tangent developable surface has a zero valued distribution parameter, while the distribution parameter of parallel rulings remains undefined. Generally, a twisted ruled surface has a non-zero distribution parameter. If adopting time t as the parameter of the dual spherical spline, it is quite easy to convert the tool path to the motion code. The velocity of the line is given by the expression:

$$\frac{d\hat{\mathbf{x}}}{dt} = \dot{\hat{\mathbf{x}}} = \hat{\omega} \times \hat{\mathbf{x}}, \tag{8.13}$$

where the angular velocity vector $\hat{\omega} = \omega \mathbf{t} + \varepsilon v \mathbf{t}$ denotes the rotation and translation along the screw axis $\hat{\mathbf{t}}$. In this local coordinate frame, an arbitrary point $\mathbf{P} = s\mathbf{r}$ along the ruling is written as $\mathbf{r} = \frac{x}{\|x\|}$, where s is the distance from the striction point $\mathbf{a}$. The velocity of this point will have a term perpendicular to $\mathbf{r}$ due to the rotation about the line $\hat{\mathbf{t}}$ and a term in the direction of $\hat{\mathbf{t}}$ due to the translation of the line. The velocity of the point can be written as [55]:

$$\begin{aligned} \mathbf{v}_p &= \omega \mathbf{t} \times s\mathbf{r} + v\mathbf{t} \\ &= -\omega s\mathbf{n} + v\mathbf{t}. \end{aligned} \tag{8.14}$$

8.4 Simulation Result

We test this approach with different turbocharger blades and simulate the manufacturing process of a blade with a cylindrical cutter. A blade consists of two sides: pressure surface and suction surface, which are manufactured respectively. The design strategies for both sides are similar. Here, we only take one example to explain the procedure.

To achieve large material removal rate, the radius of the cylinder should be large, but it must be less than the distance between two blades. Therefore, for different turbochargers, the tool sizes are varied. For this test case, the radius of the cylinder is chosen as $R = 2mm$.

The input file is a "blade.ibl" file from the software "Bladegen". It contains discrete points on different blade layers. We first extract the data for the pressure side of the blade. The offset surface is derived based on Eq. (8.1) and $d = R$.

Fig. 8.8 shows the simulation results of this approach. Fig. 8.8(a) shows the given surface and the offset surface. Then the kinematic ruled surface approximation algorithm is applied to the offset surface to get the drive surface as a ruled surface. Fig. 8.8(b) shows the discrete cutter locations which are derived after the first step of the kinematic ruled surface approximation algorithm. Fig. 8.8(c) shows the movement of the cylindrical cutter. Consequently, a surface is produced due to the movement of the cylindrical cutter. Fig. 8.8(d) compares the produced surface with the

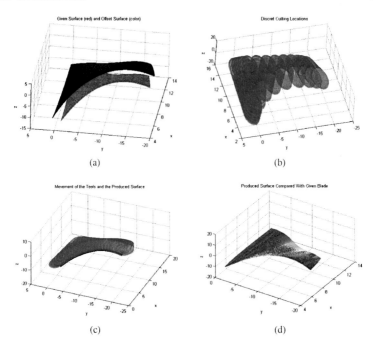

Fig. 8.8 Design a flank millable turbocharger blade: (a) Given blade surface (colored) and offset surface (red); (b) CL data; (c) Movement of the cutter and the produced surface; (d) Comparison between the given surface (colored) and the produced surface (red).

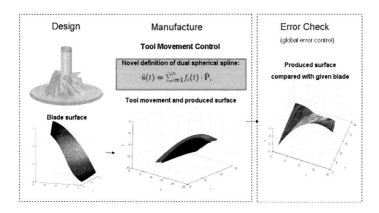

Fig. 8.9 Novel design and manufacturing procedure for turbocharger blades.

original design surface. We evaluate the error between two surfaces as the distance along z direction. The average error for these two surface is only $0.0027mm$, which is much smaller compared with the convectional tool position optimization methods. If we design the blade as a tool path which generates the blade, this blade can be manufactured accurately.

The approximation error is strongly related to the original shape of the blade. For certain cases, the approximation error is larger, especially when the original design is extremely twist and curved. For those cases, more line segments are needed to obtain a delicate approximation. Sometimes, we need to divide the blade into several regions, and apply the approximation algorithm for each region. For the piecewise ruled surface approximation, the continuity property should be treated carefully for the joint part.

Based on the approach described above, we get a novel flow chart for the turbocharger blade design and manufacturing. In Fig. 8.4, the design and manufacturing parts are combined together. This new approach avoids introducing approximation error twice and reduces the developing time.

8.5 Comments

In this chapter, we extend the kinematic ruled surface approximation in the application of the flank milling in 5-axis CNC machining combining the offset theory. Integrated the constrains of different CNC machines, it can be used as a control program to guide the movement of the manufacturing tool in flank milling process. A tool path of a cylindrical cutter is given in the form of a dual spherical spline, which describes the movement of the cylindrical cutter axis. This novel representation of tool path provides a convenient transformation to the tool motion, which leads naturally to the post-process. In addition, the tool position is calculated directly from the information of the given surface, which avoids introducing twice approximation error. A new approach for blade design is proposed, which integrates the manufacturing requirements. This algorithm can also be adapted to generate a tool path for the face milling, because the movement of the tool axis constitutes a ruled surface. For this application, the objective is to generate a tool path that is related to the normals to the surface. The manufacturing tool is not only limited to cylinder. It can be cone or even general shape. Considering the different geometry of the manufacturing tools, this algorithm has many other applications. There are still a lot of works that can be accomplished in this area.

Chapter 9
Conclusion and Future Work

9.1 Conclusion

The motivation of this dissertation is to provide an optimal design approach with ruled surface considering its manufacturing advantages. To close link the geometric design with the manufacturing process, a kinematic ruled surface approximation method is proposed. It is later extended to set up a prototype of ruled surface optimization and further applied as a control program of the flank milling process in 5-axis CNC machining. Finally, a novel design and manufacturing approach is proposed, which integrates the manufacturing requirements in the design phase and describe a desired surface as a tool path.

In Chapter 2, the theoretical backgrounds of line geometry and kinematics, which lead to different interpretations of ruled surface, are introduced. It widens the classic definition of ruled surface in Euclidean space and sets up the relation between the geometric design and manufacturing process. In particular, a ruled surface is expressed as a curve on the dual unit sphere through the Klein mapping and the Study mapping. Each point on this curve is corresponding to a ruling of ruled surface in Euclidean space. The movement of a line which generates a ruled surface is equivalent to the displacement of a point on the dual unit sphere. A ruled surface approximation problem in Euclidean space is transformed to a curve approximation problem in a higher dimensional space. This natural bonding lays theoretical basis for the proposed algorithms.

In Chapter 3, the previous work concerning ruled surface approximation is shortly reviewed and one typical approximation algorithm is implemented. This algorithm is adapted to approximate a blade surface of a turbocharger as a ruled surface. Though this algorithm does not give guides to the manufacturing process, it provides a method to pre-process the given surface and extracts line sequence for further applications.

In Chapter 4, we define a weighted average on the dual unit sphere and a dual spherical spline based on a minimization problem. The uniqueness of the definition is proven. The continuity and convexity properties of the new defined spline are discussed. It provides useful tools to represent a ruled surface and plan a tool path.

In Chapter 5, a series algorithm calculating the weighted average on dual unit sphere and interpolating the dual spherical spline are proposed. These algorithms are verified with different inputs and extended to a kinematic ruled surface approximation algorithm. The approximated ruled surface is written as a dual spherical spline which describes the movement of a line generating the ruled surface. It has direct link to the manufacturing process. Applying the inverse Klein and the inverse Study mapping, we can transform the spline back to the Euclidean space as a ruled surface.

In Chapter 6, the kinematic ruled surface approximation algorithm is improved to attain precise boundary control. It is suitable for blade surface design by approximating an original design with ruled surface. This algorithm is tested with different blade surfaces and most of them are successfully approximated as a ruled surface within an acceptable error tolerance.

Though the kinematic ruled surface approximation algorithm provides an efficient way to design a blade surface as a ruled surface, it is still a compromise between the performance and manufacturing cost. Furthermore, the original shape limits the approximated surface. In certain cases, more flexibility is required in order to design an optimal ruled surface. In Chapter 7, a prototype of ruled surface optimization is set up. In this model, a ruled surface is defined as a dual spherical spline. The control points of the spline are extracted as parameters for optimization process. By adjusting those parameters, engineers can change the shape of ruled surface. Compared with the conventional parametrization method of ruled surface, such as tensor product surface, it can reduce one third number of the parameters.

As stated before, the novel representation of ruled surface as a dual spherical spline has benefits in manufacturing phase, since it describes the trajectory of the moving line and leads to a natural representation of tool motion. Finally, the algorithms are applied to generate a tool path for 5-axis CNC machining in Chapter 8. The compact data structure of the spline provides a possibility to derive the tool orientation and location simultaneously, which is important for the 5-axis CNC machining. The smooth single parameter path representation simplified the calculation of join points corresponding the the traditional segment path representation. Using screw theory and kinematics, it is easy to judge whether the desired path is in the tool work space or not. Combining with the offset theory, we design a blade surface as a tool path of a cylindrical cutter. It can be accurately manufactured by flank milling method and avoid introducing twice manufacturing error. In this way, we integrate the manufacturing requirements in the design phase and propose a new flow chart of blade design and manufacturing.

9.2 Future Work

Many applications concerning kinematic ruled surface approximation and optimization are discussed in this dissertation. However, there are still a lot of work that can be done in the future research.

We define a dual spherical spline based on the notation of dual vectors. It will be interesting to explore the similar definition in higher dimension based on the theory of dual quaternion and define a rational dual spherical spline. This kind of spline has special applications in some industrial cases. Besides, the ruled surface optimization prototype is going to be connected to certain applications. This application needs to be examined further with an emphasis on computation efficiency.

Besides the flank milling in CNC machining, the EDM and the laser cutting, the ruled surface algorithm can be applied to other types of manufacturing operations. For some application, the location of the tool is determined by the position and orienting of its axis. Therefore, the drive surface is a ruled surface regardless of the shape of the manufacturing tools. Any manufacturing technique that depends on positioning and orienting a tool with respect to the workpiece can be modeled as a ruled surface path. These applications are worthy to be studied.

References

1. P. Alfeld, M. Neamtu, and L. Schumaker. Fitting scattered data on sphere-like surfaces using spherical splines. *Journal of Computational and Applied Mathematics*, 73(1-2):5–43, October 1996.

2. P. Alfeld, M. Neamtu, and L. L. Schumaker. Bernstein–Bézier polynomials on spheres and sphere-like surfaces. *Computer Aided Geometric Design*, 13(4):333–349, June 1996.

3. G. Arfken. *Mathematical methods for physicists, 3rd ed.* Academic Press, Orlando, FL, 1985.

4. M. Berger. *Geometry II: v.2 (Electronic Workshops in Computing), 3rd ed.* Springer, Berlin, 1987.

5. S. R. Buss and J. P. Fillmore. Spherical averages and applications to spherical splines and interpolation. *ACM Transactions on Graphics*, 20(2):95–126, 2001.

6. H. Chen and H. Pottmann. Approximation by ruled surfaces. *J. Comput. Appl. Math.*, 102(1):143–156, 1999.

7. B. K. Choi, J. W Park, and C. S Jun. Cutter-location data optimization in 5-axis surface machining. *Computer Aided Design*, 25(6):377–386, 1993.

8. C. H. Chu and J. T. Chen. Tool path planning for five-axis flank milling with developable surface approximation. *The International Journal of Advanced Manufacturing Technology*, 29(7-8):707–713, 2006.

9. W. K. Clifford. Preliminary sketch of biquaternions. In R. Tucker, editor, *Mathematical Papers*. Macmillan, 1873.

10. D. D. Dibitonto, P. T. Eubank, M. A. Petel, and M. A. Barrufet. Theoretical models of the electrical discharge machining process, I a simple cathode erosion model. *Journal of Applied Physics*, 66:4091–4103, 1989.

11. F. M. Dimentberg. *The screw calculus and its application in mechanics*. Clearinghouse for Federal Scientific and Technical Information, Springfield, Virginia, 1965. English Translation: AD680993.

12. R. Ding. Drawing ruled surfaces using the dual de boor algorithm. In *Proceedings of Computing: The Australasian Theory Symposium CATS 2002*, volume 61, pages 178–190, Melbourne, Australia, 2002.

13. R. Ding. Dual space drawing methods of cylinders. In *Proceedings of The 2003 International Conference on Computational Science and Its Applications*, page 971, Montreal, Canada, 2003.

14. R. Ding and Y. Zhang. The dual drawing method of the hyperbolic paraboloid and the screen representation of the ruling. In *Proceedings of the 2003 International Conference on Imaging Science, Systems, and Technology*, volume 2, pages 410–415, Las Vegas, Nevada, USA, 2003.

15. W. Edge. *Thoery of ruled surface*. Cambridge Univ. Press, 1931.

16. G. Farin. *Curves and surfaces for Computer Aided Geometric Design, a practical guide*. Academic Press Professional, Inc., San Diego, CA, USA, 1990.

17. S. Flöry and M. Hofer. Constrained curve fitting on manifolds. *Computer Aided Design*, 40(1):25–34, 2008.

18. H. Gong, L. X. Cao, and J. Liu. Improved positioning of cylindrical cutter for flank milling ruled surfaces. *Computer Aided Design*, 37:1205–1213, 2005.

19. J. Hoschek and D. Lasser. *Fundamentals of Computer Aided Geometric Design*. A. K. Peters, Ltd., Natick, NA, USA, 1993. Translator-L. L. Schumaker.

20. J. Hoschek and U. Schwanecke. Interpolation and approximation with ruled surfaces. In R. Cripps, editor, *The Mathematics of Surfaces VIII*, pages 213–231. Information Geometers, 1998.

21. D. Jang, J. Shin, H. Lee, S. Ro, and G. Yang. A study on tool path generation for machining impellers with 5-axis NC machine. `http://www.premalab.re.kr/seminar/seminar_data/Jang-Shin-Lee-Noh-Yang-Korea-OK.pdf`.

22. M. Kim. Intersecting surfaces of special types. In *SMI '99: Proceedings of the International Conference on Shape Modeling and Applications*, pages 122–128, Washington, DC, USA, 1999. IEEE Computer Society.

23. C. Li. *Surface design for flank milling*. PhD thesis, University of Waterloo, Ontrio, Canada, 2007.

24. X. Liu. Five-axis NC cylindrical milling of sculptured surfaces. *Computer Aided Design*, 27(12):887–894, 1995.

25. S. S. Makhanov and W. Anotaipaiboon. *Advanced numerical method to optimize cutting operations of five axis milling machines*. Springer, Berlin, 2007.

26. D. Manocha. Solving systems of polynomial equations. *IEEE Comput. Graph. Appl.*, 14(2):46–55, 1994.

27. K. Marciniak. *Geometric modeling for numerically controlled machining*. Oxford University Press, New York, USA, 1991.

28. J. M. Martínez. Practical Quasi-Newton methods for solving nonlinear systems. *J. Comput. Appl. Math.*, 124(1-2):97–121, 2000.

29. J. M. McCarthy. *Introduction to theoretical kinematics.* MIT Press, Cambridge, Massachusetts, 1990.

30. C. Menzel, S. Bedi, and S. Mann. Triple tangent flank milling of ruled surface. *Computer Aided Design*, 36(3):289–296(8), 2004.

31. E. Oberg, F. D. Jones, H. Ryffel, C. McCauley, and R. Heald. *Machiner's Handbook, 28th REV.* Industrial Press, 2008.

32. Industrial Centre of Hongkong Polytechnic University. http://mmu.ic.polyu.edu.hk/handout/0102/0102.htm.

33. S. Papageourgiou and N. Aspragathos. Rational ruled surfaces construction by interpolating dual unit vectors representing lines. In *The 14th International Conference in Central Europe on Computer Graphics, Visualization and Computer Vision*, Plzen, Czech Republic, January-February 2006.

34. N. M. Patrikalakis, T. Maekawa, K. H. Ko, and H. Mukundan. Surface to surface intersection. *IEEE Comput. Graph. Appl.*, 13(1):85–95, 1993.

35. M. A. Petel, M. A. Barrufet, P. T. Eubank, and D. D. Dibitonto. Theoretical models of the electrical discharge machining process, II the anode erosion model. *Journal of Applied Physics*, 66:4104–4111, 1989.

36. M. Peternell. G1-Hermite interpolation of ruled surfaces. *Mathematical methods for curves and surfaces: Oslo 2000*, pages 413–422, 2001.

37. M. Peternell and H. Pottmann. Interpolating functions on lines in 3-space. *The Mathematics of Surfaces VIII (R.Cripps, ed) Information Geometers*, pages 213–231, 1998.

38. M. Peternell, H. Pottmann, and B. Ravani. On the computational geometry of ruled surfaces. *Computer Aided Design*, 31:17–32, 1999.

39. L. Piegl and W. Tiller. *The NURBS Book.* Springer, 1997.

40. J. Plucker. *On a new geometry of space.* Philosophical Transactions of the Royal Society, 1965.

41. H. Pottman and S. Leopoldseder. A concept for parametric surface fitting which avoids the parametrization problem. *Computer Aided Geometric Design*, 20:343–362, 2003.

42. H. Pottmann, S. Leopoldseder, and M. Hofer. Approximation with active B-spline curves and surfaces. In *Proc. Pacific Graphics*, pages 8–25, 2002.

43. H. Pottmann, W. Lü, and B. Ravani. Rational ruled surface and their offsets. *Graphical Models and Image Processing*, 58:544–552, 1996.

44. H. Pottmann and J. Wallner. *Computational line geometry.* Mathematics and Visualization. Springer, Berlin, 2001.

45. B. Ravani and T. S. Ku. Bertrand offsets of ruled and developable surfaces. *Computer Aided Design*, 23(2):145–152, 1991.

46. B. Ravani and J. W. Wang. Computer aided geometric design of line constructs. *Transactions of the ASME Journal of Mechanical Design*, 113:363–371, 1991.

47. J. M. Redonnet, W. Rubio, and G. Dessein. Side milling of ruled surfaces: optimum positioning of the milling cutter and calculation of interference. *Advanced Manufacturing Technology*, 14:459–463, 1998.

48. F. Rehsteiner and H. J. Renker. Collision-free five-axis milling of twisted ruled surfaces. *CIRP ANNALs*, 42(1):457–461, 1993.

49. T. H. Robert, A. K. Dell, and A. Leo. *Manufacturing Processes Reference Guide*. Industrial Press, 1994.

50. J. Senatore, F. Monies, Y. Landon, and W. Rubio. Optimising positioning of the axis of a milling cutter on an offset surface by geometric error minimization. *The International Journal of Advanced Manufacturing Technology*, 37(9-10):861–871, 2008.

51. I. H. Sloan and R. S. Womersley. Constructive polynomial approximation on the sphere. *J. Approx. Theory*, 103(1):91–118, 2000.

52. K. Sprott and B. Ravani. Ruled surfaces, Lie groups and mesh generation. In *1997 ASME Design Engineering Technical Conferences*, Sacramento, California, USA, 1997.

53. K. Sprott and B. Ravani. Kinematic generation of ruled surface. *Advances in Computational Mathematics*, 17:115–133, 2001.

54. K. Sprott and B. Ravani. Cylindrical milling of ruled surface. *The International Journal of Advanced Manufacturing*, 38(7-82):649–656, 2007.

55. K. S. Sprott. *Kinematically generated ruled surfaces with applications in NC maching*. PhD thesis, University of California, Davis, 2000.

56. J. Stillwell. *Naive Lie Theory*. Springer, Berlin, 2008.

57. Hongkong Polytechnic University. `http://mmu.ic.polyu.edu.hk/handout/0103/0103.htm`.

58. J. V. Valentino and J. Goldenberg. *Introduction to Computer Numerical Control (CNC) (3rd Edition)*. Prentice Hall, 2002.

59. G. R. Veldkamp. On the use of dual numbers, vectors and matrices in instantaneous, spatial kinematics. *Mechanism and Machine Theory*, 11:141–156, 1976.

60. X. Wang, W. Zhang, and L. Zhang. Intersection of a ruled surface with a free-form surface. *Numerical Algorithm*, 46:85–100, 2007.

61. Wikipedia. `http://en.wikipedia.org/wiki/Turbocharger`.

62. J. Xia and Q. J. Ge. Kinematic approximation of ruled surfaces using NURBS motions of a cylindrical cutter. In *Proceedings of ASME 2000 Design Engineering Technical Conferences*, pages 1–7, Baltimore, Maryland, USA, 2000.

63. Y. Xu. Polynomial interpolation on the unit sphere and on the unit ball. *Advances in Computational Mathematics*, 20(1-3):247–260, January 2004.

64. X. F. Zha. A new approach to generation of ruled surfaces and its applications in engineering. *The International Journal of Advanced Manufacturing Technology*, 13:155–163, 1997.

65. Y. Zhou, J. Schulze, and S. Schäffler. Flank millable blade design for centrifugal compressor. In *Proceedings of the Mediterranean Conference on Control and Automation*, pages 646–650, Los Alamito, CA, USA, June 2009. IEEE Computer Society.

66. Y. Zhou, J. Schulze, and S. Schäffler. Blade geometry design with kinematic ruled surface approximation. In *SAC '10: Proceedings of the 2010 ACM Symposium on Applied Computing*, pages 1266–1267, New York, NY, USA, 2010. ACM.

Curriculum Vitae

Name: Yayun Zhou

Nationality: China

Date of Birth: 10 June, 1981

Place of Birth: Hangzhou, China

Education:

- **2007 - 2010:** Department of Electronics and Information Technology,

    Universität der Bundeswehr München, Germany

- **2006 - 2007:** Industrial Mathematics Institute,

    Johannes Kepler University Linz, Austria

- **2005 - 2006:** Department of Computer Science and Mathematics,

    Technology University Eindhoven, Netherlands

- **1999 - 2005:** Department of Information and Electronics Engineering,

    Zhejiang University, China

Awards and Honors:

- **2009 - 2010:** Ernst von Siemens Promotionsstipendium

- **2005 - 2007:** Erasmus Mundus Scholarship from European Commission

- **2002 - 2003:** Outstanding Graduate of Zhejiang Province